Cosmology:
The History and Nature
of Our Universe
Part II

Professor Mark Whittle

THE TEACHING COMPANY ®

PUBLISHED BY:

THE TEACHING COMPANY
4840 Westfields Boulevard, Suite 500
Chantilly, Virginia 20151-2299
1-800-TEACH-12
Fax—703-378-3819
www.teach12.com

ISBN 1-59803-494-4

Mark Whittle, Ph.D.

Professor of Astronomy, University of Virginia

Professor Mark Whittle has been a member of the astronomy department at the University of Virginia since 1986. He was a Mackinnon Scholar at Magdalen College, University of Oxford, and obtained a First-Class Honors degree in Physics in 1977. He received his Ph.D. in Astronomy from the Institute of Astronomy at the University of Cambridge in 1982. That same year, he received NATO and Lindemann postdoctoral fellowships at Steward Observatory at the University of Arizona. From 1984 to 1986, he was a Research Fellow at Jesus College, University of Cambridge.

Professor Whittle has published more than 80 articles in research journals. Two of the more important ones are "Virial and Jet-Induced Velocities in Seyfert Galaxies" (*Astrophysical Journal*, vol. 387) and "Jet-Gas Interactions in Markarian 78" (*Astronomical Journal*, vol. 129). Professor Whittle's research interests center on various aspects of nuclear activity in galaxies, a phenomenon that arises from gas falling onto supermassive black holes. His most recent work focuses on the properties of jets that emerge from Seyfert galaxies and the role they play in energizing the central few thousand light-years. He uses both optical and radio telescopes for his research, including the Hubble Space Telescope.

Professor Whittle has been a regular user of national and international observatories, and he has served on time-allocation panels for the Hubble Space Telescope, the Chandra X-ray Telescope, and the Kitt Peak National Observatory, as well as on several grant-allocation panels for the National Science Foundation. He is a member of the Royal Astronomical Society, the American Astronomical Society, and the International Astronomical Union.

Professor Whittle has been teaching courses for nearly 20 years at both the introductory and graduate levels, including courses on the solar system, stars, galaxies, cosmology, and observing techniques. He was awarded the Harrison Prize for Outstanding Academic Advising in 1999. Within the astronomy department, he was chair of the undergraduate program from 1991 to 2005, and he is currently chair of graduate admissions. He gives frequent public lectures, and he is committed to community outreach on the subject of Big Bang acoustics.

Table of Contents

Cosmology: The History and Nature of Our Universe Part II

Cosmology: The History and Nature of Our Universe

Scope:

Cosmology is the study of the origin and nature of the entire Universe. It is currently one of the most exciting of the academic disciplines, caught in the midst of a truly remarkable period of growth. After almost a century of slow but steady progress, in the last decade, several exceptional observational and theoretical projects have really transformed the subject, removing previous confusion and uncovering a story of stunning richness and coherence. The progress has been so significant that there is an almost palpable feeling of excitement within the astronomical community. With all these recent developments, and with this new level of clarity, now is the perfect time to explore cosmology.

These 36 lectures are at a level that assumes no prior knowledge of astronomy, nor science beyond high school, although they will take us well beyond more traditional astronomy survey courses. Despite its apparently esoteric nature, much of the cosmological story is directly accessible to our intuition and can be told in everyday language.

The lecture series moves through eight themes that combine to give a coherent picture of the entire subject. After an introductory lecture that sets the stage, we devote 8 lectures to an overview of the Universe. We review its overall form and introduce its hierarchy of structures, from atoms to stars to galaxies to the cosmic web. We identify its fundamental constituents and outline its overall history from the Big Bang to today. We explore how distances are measured and how these lead to the discovery of cosmic expansion, along with our ability to witness all of cosmic history directly by simply looking far away. Finally, we get acquainted with the Universe's two most unusual constituents, dark matter and dark energy.

Our second theme takes 3 lectures to explore the geometry of cosmic space and its expansion history. This introduces us to the gravitational theories of both Newton and Einstein, and it frames the expansion as a delicate balance between the kinetic energy of expansion and the gravitational energy acting between the Universe's constituents. We'll explore how each cosmic constituent affects the expansion, taking time to understand dark energy's remarkable tendency to "fall outward," leading to an accelerating expansion.

In our third theme we take 4 lectures to visit the first million years. This era is revealed to us directly and in considerable detail by the cosmic microwave background: an extremely distant wall of brightly glowing fog that is, roughly speaking, the actual fireball of the Hot Big Bang. The most recent development has been to detect and study primordial sound waves pulsing through the hot gas. The nature of this sound (which we will listen to) reveals much about the properties of the Universe, and it is one of the primary reasons for such rapid recent progress in cosmology.

Our fourth theme takes 6 lectures to pursue the subsequent evolution of the sound waves as they grow to form the first stars, the first galaxies, and even to leave their mark on the weblike pattern of thousands of galaxies that span the night sky. We'll see how dark matter plays a critical role, gathering the expanding cooling gas into ever-larger structures, and we'll follow the evolution of galaxies across all of cosmic history, tracing the processes that allow or suppress their formation and growth. We also step back to watch the development of the large-scale weblike patterns of galaxies, and we learn that these patterns are just now becoming frozen as dark energy's accelerating expansion inhibits any further growth.

Our fifth theme takes 4 lectures to trace the origin of the chemical elements—without which you and I could not be made. Atom factories, we'll learn, lie deep at the centers of stars. When stars die, their newly made atoms are brought to the surface and blown off into space, where they can later form planets and people. We'll push the subject to a higher level and explain the origin of the complex pattern of element abundances, from carbon to uranium, and we'll visit the minute-old Universe, where the lightest elements were forged in the heat of the Big Bang. Finally, we'll see how cosmologists use 5 major data sets to measure many of the Universe's basic properties, arriving at a "concordance cosmology" with remarkable precision.

Our sixth theme takes us back to the very early Universe, when it was less than a second old. In 3 lectures we first develop a feel for matter at extremes of temperature and density, and we then use that understanding to view the cascade of changing particles that emerges as the Universe expands and cools, including several moments of immense carnage as particles and their antiparticles annihilate. We'll also learn how the forces of Nature, and even the laws of physics, emerged and evolved within the first second. This is a realm in which several profound questions remain unanswered, although exciting breakthroughs are anticipated in the near future.

Our seventh theme devotes 4 lectures to the remarkable process of inflation—the launch for the expansion, the "Bang" of the Big Bang. We begin by learning about several puzzling problems with the Standard Big Bang Theory and then explore how all these problems can be solved if the Universe began with a brief period of accelerated expansion. What might drive such an expansion? An extreme version of present-day dark energy—perhaps a large energy associated with the vacuum of space—might cause it to fall outward rapidly. In this kind of expansion, tiny quantum fluctuations are amplified and become the seeds for star and galaxy formation. This is a truly amazing concept: stars, galaxies, and the galaxy web owe their existence to quantum processes way down in the subatomic microworld. But inflation has other stunning qualities: It can create the Universe from almost nothing, balancing arbitrarily large amounts of positive matter and energy against equal amounts of negative gravitational energy. Finally, not only can it create a Universe vastly bigger than our visible portion, it may also create multiple Universes, each with its own laws of physics.

Our final theme returns home and aims to place *us* into the cosmological story. We first look at a deeply puzzling property of the Universe: its amazing fertility. Why does the Universe seem "optimized" for life? There are several answers, although at this time none are widely accepted. We'll also look at the likely progress in cosmology in the next couple of decades. The subject is in a healthy state of growth, and there are many exciting projects to look forward to. Finally, we end the course by considering two philosophical questions: Why is the remote Universe so easily comprehensible to us even though our minds were, presumably, designed for quite different purposes? And can modern cosmology play the same role in society that traditional creation stories have done throughout history—namely, provide an emotionally and spiritually nurturing framework within which we can find a broader meaning for our lives and our culture?

Cosmology is a wonderful, profound, and fascinating subject, and given its remarkable recent progress, now is the perfect time to learn about it.

Lecture Thirteen
The Cosmic Microwave Background

Scope:

By looking sufficiently far away (hence, back in time) we can actually witness the "flash" from the Big Bang itself. Since all directions point back in time, this flash is seen as an omnidirectional glow. Because of cosmic expansion and its associated red shift, what was a brilliant flash of light arrives as a rather feeble glow of microwave radiation, called the cosmic microwave background (CMB). We'll review how the CMB was discovered in the early 1960s and its subsequent scrutiny over the last 50 years. Modern observations reveal the CMB to be exceedingly uniform, with very slight patchiness. The former reveals the smoothness of the young Universe, while the latter reveals primordial sound waves that are the seeds from which stars and galaxies will later grow. We'll discuss how the CMB was formed and how it gives us a direct window on the preembryonic Universe.

Outline

I. This lecture begins Theme III: The First Million Years.

   A. It begins our steady march through cosmological history.

   B. The first million years is a great place to start.

      1. It gives us a window on the even-earlier Universe, and it sets the stage for the birth of the first stars and galaxies.

      2. It's the earliest time we can witness directly, via the microwave background, so it's on a secure observational footing.

      3. The microwave background provides a gold mine of information.

      4. The conditions are relatively easy to understand and intuit, and in human terms they are stunningly impressive.

II. Recall from Lecture Two that as we look out into the Universe, we look back in time.

   A. If we look 14 billion light-years away, we look 14 billion years ago, so we should see the Big Bang.

B. Because all directions point back in time, we're surrounded by a remote 360° panorama of the Universe's birth.

C. The young Universe was 3000 K, so one might expect to see a bright orange glow all around us. In fact, red shift stretches the light waves into the microwave part of the spectrum, invisible to our eyes.

D. The cosmic microwave background (CMB) is our view of the young Universe.

III. Let's clarify what the microwave background actually is.

    A. It's a view of the Universe 400,000 years *after* the Big Bang.

        1. The young Universe is so hot it's ionized, and so the gas is foggy.

        2. Expansion cooled the Universe, and after 400,000 years the temperature passed 3000 K, at which point atoms formed and the fog cleared.

        3. So, looking outward (and back in time), our vision is limited by this bright, glowing wall of fog.

        4. As the light crosses the expanding Universe, its waves are stretched 1000-fold and arrive as microwaves.

    B. The microwave background reveals a "preembryonic" Universe.

        1. Relative to a human lifespan, the CMB is 12 hours after conception.

        2. Relative to a 26-mile marathon, the CMB is 4 feet from the finish.

IV. Let's look at the prediction and discovery of the microwave background.

    A. In the mid-1940s, George Gamow wondered whether a Hot Big Bang might create all the chemical elements.

        1. He predicted that thermal emission from the fireball should still be visible as microwave radiation.

        2. Current instruments weren't very sensitive, so no one looked for it.

        3. The prediction was forgotten, in part because the element creation idea didn't work.

B. In the 1960s, two groups independently recalculated the properties of a Hot Big Bang: Yakov Zeldovich and collaborators in Moscow, and Jim Peebles and Robert Dicke at Princeton. Both groups predicted a microwave background, and Dicke began building a microwave detector.

C. Meanwhile, as part of a NASA program, Arno Penzias and Robert Wilson were making sensitive microwave observations at Bell Labs.

    1. They could not understand the origin of 55% of their signal.

    2. Penzias heard about the Princeton group's Big Bang prediction and realized that was the origin of the missing signal.

    3. Penzias and Wilson received a Nobel Prize for discovering the cosmic microwave background.

D. In retrospect, the background radiation had already been detected on several occasions by different groups but had been dismissed as too uncertain, or its significance hadn't been appreciated.

E. This topic nicely illustrates the often-crooked path that scientific progress can take.

V. Since its discovery, the microwave background has been the focus of enormous scientific scrutiny, both observational and theoretical.

A. The most impressive results have come from satellites, including NASA's Cosmic Background Explorer (COBE) and the Wilkinson Microwave Anisotropy Probe (WMAP).

B. In 1990, the COBE satellite made a precise measurement of the CMB spectrum.

    1. It is an amazingly accurate thermal spectrum, with a temperature of 2.725 ± 0.002 K—very cold!

    2. This makes sense: Applying a red shift of 1000 to a 3000 K spectrum turns it into a 3 K spectrum.

    3. Alternatively, cosmic expansion has cooled the Universe from 3000 K to 3 K, which is simply the current temperature of the Universe.

   C. The accurate thermal shape also confirms the following facts.

     1. The Big Bang was hot, not cold—hence, the "Hot Big Bang."

     2. High-temperature physics applies to the early Universe.

     3. The early Universe was in a particularly simple state called "thermodynamic equilibrium."

VI. With a 360° panorama, what image does the CMB show?

   A. The answer depends on the degree of contrast enhancement. At the lowest level, the CMB appears extremely uniform.

     1. This is the best demonstration that the Universe is isotropic.

     2. It tells us that the young Universe is very smooth, just uniform glowing gas. This is fortunate because smooth gases are much easier to study than clumpy gases.

   B. Contrast stretching by about 1000 reveals a dipole (yin-yang) pattern, brighter on one side of the sky and dimmer on the other.

     1. This arises from a Doppler shift caused by our galaxy moving through space at 600 km/s. The CMB is brighter ahead and dimmer behind.

     2. We're "falling" toward a large collection of galaxies a few hundred million light-years away, in the direction of the constellation Hydra.

   C. After removing the dipole pattern and contrast stretching again, we see emission from the Milky Way. Fortunately, it's possible to remove this emission because it has a different microwave "color" than the CMB.

   D. The final CMB image reveals a mottled pattern of slightly brighter and darker patches. The amplitude of the patches is only 1/100,000: the height of a bacterium on a bowling ball. We can think of these patches in two ways.

     1. They are regions of slightly higher and lower *density* that will ultimately become galaxies and voids—they're the seeds of galaxies.

     2. They are regions of slightly higher and lower *pressure*— they're peaks and troughs of primordial sound waves.

E. The patchiness contains a huge amount of information that we can illustrate with a metaphor.

　　1. Twelve hours after conception, a human embryo has no structures; it's just a single cell with its DNA. Likewise, at the time of the CMB, the Universe has no structures; it's just a uniform gas filled with sound waves.

　　2. DNA contains much information about our ancestral past, and it determines in large part our future development. So it is with cosmic sound waves and the Universe.

　　3. In a sense, studying the CMB and its patchiness is to astronomy what the Human Genome Project is to the life sciences. One might call it the "cosmic genome project."

Suggested Reading:

Hawley and Holcomb, *Foundations of Modern Cosmology*, chap. 14.

Mather and Boslough, *The Very First Light*.

Smoot and Davidson, *Wrinkles in Time*.

Lemonick, *Echo of the Big Bang*.

Bartusiak, *Archives of the Universe*, chaps. 45, 63 (prediction and discovery papers).

Questions to Consider:

1. Explain the remarkable fact that we can "see," all around us, the light from the Big Bang. More specifically, how can we see something that ended 14 billion years ago? Why do we not see the Big Bang itself, but instead a time 400,000 years afterward? What was the color of the light emitted at that time, and why do we now see it as microwaves?

2. Explain the main features of the microwave background radiation. What spectrum does it have, and why? Why is it so smooth? What causes the dipole pattern and band around the sky, and how are these removed from the data? What causes the very slight patches, and why do these patches contain so much information?

3. Summarize the history of the prediction and discovery of the cosmic microwave background. Review the more recent instruments that have given much more detailed information on the CMB. What instruments are planned for the future, and what are they aiming to measure?

4. Using the fact that the microwave background was formed roughly 400,000 years after the Big Bang, which itself occurred about 14 billion years ago, confirm the examples given in the lecture that this is equivalent to 12 hours after conception in an 80-year human lifespan, or 4 feet from the finishing tape of a 26-mile marathon.

Lecture Thirteen—Transcript
The Comic Microwave Background

This lecture begins Theme III, where we'll be visiting the first million years in the Universe's life. It also begins our steady march through the various periods in cosmological history, so I want to take a few moments to get our bearings and see what the next 24 lectures and six themes will bring us, using the timeline from Lecture Five.

So here we are at Theme III. Notice we're starting neither at the beginning nor at the end, but somewhere in the middle, at this crucial transition period within the first million years. It gives us a window on the even-younger Universe, but it also sets the stage for the birth of the first stars and galaxies. So Theme IV then picks up that story and traces the evolution of stars and galaxies from their birth to the present time. Theme V looks at an important piece of our own lineage, the construction of the many types of atoms that make up you and me, and part of that story goes back to the minute-old Universe. And that prepares us to visit, in Theme VI, the extremely young nanosecond-old Universe, when matter and forces first took on their current form.

Now, up to this point, the story is remarkable in its coherence and completeness, but it lacks a launching mechanism. It lacks a theory for the bang in the Big Bang, so Theme VII will explore that theory. It's called "inflation," and it has some wonderful and extraordinary consequences. And finally, as with any cosmological story, scientific or traditional, our own conscious presence raises broader, more philosophical questions that I'll briefly explore in Theme VIII. And that'll naturally bring this course to its close.

So, let's now begin our visit to the first million years. It's a great place to start for a number of reasons. It's the earliest time we can witness the Universe directly via the microwave background, so it's on a secure observational footing. But it's also on a secure theoretical footing, because the conditions at this time are relatively simple and well understood. The microwave background also contains an absolute gold mine of information, partly about the Universe as a whole, partly about the stars and galaxies that are soon to be born, and partly about the even-younger Universe, including the inflationary period. And lastly, from our perspective, a nice quality of the first million years is that the conditions are relatively easy to understand and intuit, as well as being stunningly impressive.

As we'll see, this time is characterized by brilliantly colored skies, intense heat and loud semimusical sound, all of which we'll meet in due course.

So let's begin relatively gently by recalling that as we look out into the Universe, we also look back in time. So for a galaxy 5 billion light-years away, we see it as it was 5 billion years ago, not as it is now. So pushing this idea to its limit, if we look 14 billion light-years away, we see the Universe as it was 14 billion years ago. Well, what was going on 14 billion years ago? The Universe was being born, so that's what we see, the Universe's birth. Perhaps even more bizarrely, it doesn't matter which direction we look at—all directions end far away in the Universe's birth. So the amazing fact is that out there, beyond the stars and beyond the galaxies, lies an all-sky panorama of the Big Bang. It's just sitting there, waiting for us to study it.

Using our picture of the astronomer's view of the Universe, we can now ask, "What do we see, if anything, of this outermost shell? Is the Big Bang visible?" The answer is a very definite "yes," and it's not difficult to see why. The Big Bang was very hot, and hot things glow brightly. In fact, for reasons we'll get to in a moment, the relevant temperature was 3000 K, and we know from Lecture Four that a gas of this temperature glows with a powerful deep-orange light. So we expect to see, in every direction, a brilliant orange glow covering the sky. Well, this flies in the face of common sense, because the night sky is dark black, not bright orange. So something's clearly missing in our thinking, and what's missing is red shift.

At such a huge distance, the red shift is enormous, and the orange light waves are stretched a thousand-fold, from a millionth of a meter to a thousandth of a meter—that's a millimeter. And that's in the microwave part of the electromagnetic spectrum. So the Big Bang is directly available for us to study, as a panoramic microwave glow. This, of course, is the famous cosmic microwave background, or CMB for short. I never cease to be impressed by this. There are, in fact, as many microwave photons coming from the sky as there are photons of light coming from a full Moon, so if by some quirk of evolution we developed microwave-sensitive eyes, at night we could see our surroundings, and we would cast shadows by the light of creation. Now, what an absolutely stunning notion! So when you next look out at the night sky, try to feel the presence of the microwave background, the glow from the Big Bang. It really is there, but your eyes don't see it because of its huge red shift.

Now, before we proceed, let's clarify exactly what the microwave background actually is. For example, is it the Big Bang itself? The answer is "no," it isn't. It's the Universe 400,000 years after the Big Bang, and to see why this is, we need to look more closely at this outer portion of our astronomer's view of the Universe. So furthest away, longest ago, sits the Big Bang itself. It's like the first second, if you want. Inward and to the right are forward in time. The Universe starts out very hot—so hot that atoms can't yet form—and the gas is ionized. Nuclei and electrons separately zoom around freely. This kind of gas is foggy. You can't see through it, so the early Universe is filled with a bright, glowing fog. But as the years pass, the Universe expands and cools, and after about 400,000 years, its temperature dropped past a critical value of 3000 K, below which electrons become attached to nuclei, and atoms could finally form.

Now, this transition also makes the fog clear, and since then the Universe has been transparent. So we look out through the transparent Universe, up to, but no further than, this wall of brightly glowing fog. As the bright orange light crosses the Universe on its way to us, its waves are stretched a thousandfold by cosmic expansion, and they arrive as microwaves. So the microwave background comes from a wall of 3000 K glowing fog, 400,000 years after the Big Bang.

Now, the thickness of this boundary in this diagram is a little bit misleading. Here's a more accurate diagram of cosmic history, with the Big Bang on the left, and now on the right, 14 billion years later. If we match this to an 80-year-old human lifespan, then the Hubble telescope looks back to 10-year-old preteens. The James Webb Space Telescope will look back to 2-year-old children, but look where the microwave background is. At 400,000 years, it is equivalent to just 12 hours after conception. This isn't just the infant Universe, nor is it even the embryonic Universe—it's the preembryonic Universe. And just as for humans, the Universe at this time is radically different from the Universe we know today.

Now, let's recast this using distance instead of time. Here's a 26-mile marathon race from us to the Big Bang. In this case, the microwave background is just 4 feet from the finishing tape. How frustrating.

We can see almost to the Big Bang itself, but not quite. It's forever hidden behind an utterly impenetrable wall of fog.

Let's now spend a few minutes looking at, first, the prediction and then, the discovery of the microwave background, because it provides a nice example of the often-confused and serendipitous nature of scientific progress. The story begins in the mid-1940s, when the theoretician George Gamow began to calculate what actually happens during the early phase of the Big Bang. Gamow wanted to see if the Big Bang could make all the different kinds of atoms we find around us, the chemical elements. Now, because you need over a billion degrees to make atomic nuclei, Gamow guessed that the Big Bang might have been a very hot kind-of fireball [and] that the thermal radiation from this fireball might still be visible today but redshifted into the microwave range.

Now, at that time, the instruments were nowhere near sensitive enough to detect this radiation, so no one looked for it. zAnd in addition, a year or 2 later, more detailed calculations by other people showed that most of the chemical elements are, in fact, made in stars, not in the Big Bang. So perhaps surprisingly, this early 1940s work by Gamow, which actually predicted the existence of a microwave background, essentially disappeared and was forgotten for about 15 years. It wasn't until the early 1960s that calculations of a Hot Big Bang were done again, independently, by two groups: Yakov Zeldovich and collaborators in Moscow, and Jim Peebles and Robert Dicke at Princeton. As before, these calculations predicted the fireball's thermal radiation should be visible as redshifted microwaves.

Now, Robert Dicke was also a superb experimentalist, and so he immediately set about building a detector to see if he could see this microwave emission. And by May 1964, the Princeton group was close to making its first measurements. Now, meanwhile, completely unknown to Dicke, an hour away at Bell Telephone Labs in Holmdel, New Jersey, two radio astronomers, Arno Penzias and Robert Wilson, were also trying to make an extremely sensitive microwave receiver, but for quite different reasons. This was the early days of satellite telecommunications, and to support NASA's ECHO project, Bell Labs had built a large microwave horn antenna. And by 1964, Penzias and Wilson had managed to make an accurate measurement of the microwave emission from the sky.

But something was quite puzzling. Their signal was too strong: 35% came from the atmosphere; 10% came from their amplifier—leaving 55% unaccounted for. It came from all directions and was constant over time, and they were completely unable to explain it. Then one day, Penzias heard about the Princeton group's prediction of the microwave background emission from the Big Bang, and he immediately phoned Dicke at Princeton, who happened to be in the middle of his weekly lunch meeting with his group. At the end of the conversation, Dicke put down the phone and famously said, "Well, boys, we've been scooped." And, indeed, they had. It was Penzias and Wilson who later received a Nobel Prize for discovering the cosmic microwave background, arguably the most important scientific discovery of the century, with the possible exception of Hubble's discovery of cosmic expansion.

Now, I'd like to add that in retrospect, the background radiation had already been detected on several occasions by various different groups, but each time, it had either been sort of dismissed as too uncertain or its significance hadn't been appreciated. In fact, this whole episode in the history of science is full of missed opportunities and confusion, and it reminds us that science rarely proceeds in a clean linear fashion, especially in the midst of an unanticipated breakthrough like the discovery of the microwave background. Now, as you might imagine, in the half century since its discovery, the microwave background has been the focus of enormous scientific scrutiny, both observational and theoretical.

So let's now leave those early days and come forward to look briefly at some of the more recent observations. Here's a selection of CMB instruments from the last decade. Two are high in the Andes, and since our own atmosphere emits at least some microwaves, it's important to get above as much of it as possible, and of course, the best place to go is above the entire atmosphere, into orbit. And the most impressive results have come from satellites, including NASA's Cosmic Background Explorer, or COBE, and the Wilkinson Microwave Anisotropy Probe, or WMAP. It's named after David Wilkinson, whose long career in CMB studies began back in 1965, as one of the boys in the Bob Dicke group scooped by Penzias and Wilson.

Now, to show you how far technology has come in 50 years, WMAP was placed well beyond the Moon's orbit in a stable gravitational pocket called "L2." It took data continuously for 5 years and generated an all-sky map, about a million times more sensitive than those first detectors back in the 1960s. So let's now look at the kind of data these satellites have provided. As with any source of radiation, one of the first things astronomers ask is, "What is its spectrum? How does its brightness change with wavelength?" The reason is that a spectrum can tell you a huge amount about the source, and in this case, the source is the young Universe.

So, in 1990, the COBE satellite made an extremely accurate measurement of the CMB spectrum. Now, remember from Lecture Four, we are expecting some kind of thermal spectrum that is generated by the ceaseless collisions of nuclei and electrons in a hot gas. The Sun, for example, has an approximately thermal spectrum since its light comes from a hot atmosphere. So here are the COBE data points, spanning wavelengths from 0.5 millimeters to 5.0 millimeters. Now, through these data points is plotted a theoretical pure thermal spectrum, which as you can see, matches the data stunningly well.

I was actually present when this plot was first shown publicly at the January 1990 meeting of the American Astronomical Society in Washington, and the audience of almost 1000 astronomers spontaneously stood up and applauded for almost a minute. The applause was, I think, really for two things. They were, of course, acknowledging the incredibly hard work and skill of the COBE scientists, but they were also expressing their delight, and perhaps relief, that the Universe conforms to simple physics and might therefore be ultimately understandable.

So what do we learn from this thermal spectrum? Recall from Lecture Four, for a thermal spectrum, the peak wavelength and overall intensity depend on the temperature of the object. So what's the temperature of the microwave background? It's a very precise 2.725 ± 0.002 K. Now, remember, degrees Kelvin are degrees centigrade above absolute zero, so 2.7 K is extremely cold, and this may strike you as odd, since the gas emitting it is so hot. But don't forget red shift. When a thermal spectrum is redshifted, it stays a thermal spectrum, and the shift to longer wavelengths corresponds exactly to a shift to lower temperature. So the microwave background redshift stretch factor of 1000 takes a 3000 K thermal spectrum and turns it into a 3 K thermal spectrum.

Now, there is, in fact, another way to look at this. As the Universe expands, its contents cool, and in the 14 billion years since the microwave background was formed, the Universe has cooled from 3000 K to 3 K. And so the microwave background simply reflects the current temperature of the Universe.

Now, what else do we learn from the CMB spectrum? Well, the fact that it has a thermal shape immediately tells us that the young Universe was hot, not cold, and hence cosmologists talk not just of the Big Bang, but of the Hot Big Bang. And this in turn means we can study the very early Universe using our knowledge of high-temperature physics, as we'll see in Lecture Twenty-Seven. But perhaps more importantly, the precision of the thermal shape tells us that the young Universe was in a particularly simple state, which physicists call "thermodynamic equilibrium." In this state, all the various components—the light, the electrons, the nuclei—all exchange energy back and forth freely, and each has the same well-defined temperature.

And this enormously simplifies the situation. To a physicist, hot gas in thermodynamic equilibrium is one of the easiest things to understand. This is sometimes hard for people to accept. One somehow expects the early Universe to be desperately complex and difficult to understand, perhaps because it's such a remote and alien realm, whereas in truth, it's much simpler to understand than almost everything that we find around us here on Earth.

Let's now turn to the spatial properties of the microwave background. Since it's a 360° panorama, what image does it show? What patterns are visible, if any? Now, the first thing to realize is that what one sees depends very much on the degree of contrast enhancement. Just as normal photos look quite different as one fiddles with the contrast, so the microwave background looks quite different as one changes its contrast.

Secondly, remember that all-sky maps appear as ovals, a bit like the full Earth maps. So at the lowest level of contrast, the microwave background looks completely uniform. Remember, in Lecture Three, this is the most spectacular demonstration of isotropy. The Universe is amazingly similar in every direction. It also tells us that the young Universe is very smooth. So long before there were any stars or galaxies, the Universe was simply an infinite ocean of extremely uniform, hot glowing gas. This is also good news for cosmology because smooth gases are much easier to study than clumpy gases.

If you now subtract the average brightness and contrast stretch by about 1000, you see this odd pattern. It's a bit like a yin-yang symbol, whose technical name is a "dipole," meaning there are two poles. There is a slightly brighter area here, and a slightly dimmer area here, with a smooth change in between the two. The effect is rather interesting and is caused by the motion of our galaxy, the Milky Way, through space. If our galaxy is moving through the Universe, then the microwave background is slightly Doppler-shifted. It's bluer and brighter in the direction of motion, and it's redder and dimmer in the opposite direction. And these in turn correspond to slightly hotter and colder temperature ahead and behind. And the difference in temperature, $\Delta t/t$, gives us the velocity relative to the speed of light, V/c.

Now, using this, it turns out that our galaxy is falling at 600 kilometers per second towards the constellation of Hydra in the Southern Hemisphere. Notice I use this word, "falling." What's happening is that although galaxies basically move with the overall Hubble expansion, they're also being pulled by their neighbor galaxies, and this adds a falling motion on top of the expansion motion, and that's what makes the CMB dipole. What's causing our galaxy's motion is a huge assembly of thousands of galaxies, sitting several hundred million light-years away, in the direction of Hydra. And we're effectively falling towards it at about 600 kilometers per second. Now, don't worry if you didn't quite follow that—we'll come back to this whole topic in Lecture Twenty-Two.

Okay, now, when we subtract the giant dipole pattern and contrast stretch again, there is a strong feature running across the sky caused by the microwave emission from our own Milky Way galaxy. Now, fortunately, as this spectrum shows, the Milky Way emission has a quite different spectral shape than the microwave background's thermal emission. So as long as we have maps made at several different wavelengths, as all these WMAP images are, then you can remove this contaminating Milky Way emission to leave the pure microwave background image, shown here in the middle.

Now, this final microwave background image is extremely important in cosmology, but before looking at why, I just want to quickly review what we've done, using a spherical depiction of the sky instead of an oval one. Here's a picture of the Earth, and from it we look out into space and see the optical celestial sphere, with its stars and constellations. If we look at the same celestial sphere but using a microwave telescope, instead of seeing stars we see an amazingly uniform microwave glow, which is shown here in green. If we then remove the subtle Doppler dipole and Milky Way emission and contrast stretch by about 100,000 times, this is what we see: slightly brighter and darker patches that cover the sky. We can see them all if we spin it around like this.

Now, this pattern is made from slight patches, from small to large, all randomly superposed on top of each other. The amplitude of the patchiness is about 1 in 100,000, and that's equivalent to the height of a bacterium on a bowling ball, so this patchiness is very, very slight indeed. Now, one can view these patches in two different ways, both of which are correct. First of all, they represent regions of slightly higher and lower density in the hot gas. The over-dense regions contain slightly more mass, and gravity will ultimately pull them into galaxies and clusters of galaxies, while the slightly under-dense regions contain slightly less mass, and these will ultimately become the spaces or voids between galaxies. So the patches are the seeds from which galaxies will grow, and we'll pursue this perspective in Lectures Seventeen and Eighteen.

Now, the second way to view the brighter and darker patches is as the peaks and troughs of sound waves. You see, the higher and lower density, together with higher and lower temperature, makes higher and lower pressure in the hot gas, but pressure variations are, in fact, sound waves. So the microwave background gives us a freeze-frame image of sound waves in the early Universe. They're caught as they cross this wall of fog, and we'll pursue this perspective in Lectures Fifteen and Sixteen.

Now, it may not be obvious to you as you look at this image, but this patchiness contains a huge amount of information—what, exactly, we'll be talking about in the next couple of lectures. But to illustrate this richness of information, I'd like to end by briefly developing a metaphor that parallels the time of the microwave background with the equivalent time in a human's life, just 12 hours after conception. Now, in the case of a human, at that time, there are no structures, no arms or head or organs. There's just a single cell with its DNA. Now, likewise, in the Universe at that time, there are no structures. There are no stars or galaxies or clusters. There is just uniform hot gas, filled with sound waves.

Now, for a human, that first DNA is immensely important. It determines, to a large extent, the future development of the child and adult, and it also contains a lot of ancestral evolutionary information. So it is with the cosmic sound. It determines in large part how the Universe of stars and galaxies will grow, and it also contains information coming from much earlier times. So in a sense, then, studying the microwave background and its patchiness is to astronomy what the Human Genome Project is to the life sciences. In fact, one might almost call it the "cosmic genome project."

Well, with this introduction now in place, we're ready to visit the first million years. What was it like back then? What would we have seen and felt and heard? And why were there sound waves, and what can we learn about the Universe by listening to them? Well, we'll find out in the next three lectures.

Lecture Fourteen
Conditions during the First Million Years

Scope:

What was the Universe like during its first million years? We'll answer this question using both physical terms as well as human experiential terms. Everywhere was an exceedingly hot, thin, glowing "atmosphere" of hydrogen and helium. Although the gas itself was too thin to burn you, light's brilliance would kill you instantly. As time passed and expansion caused the temperature to drop, the color of the sky slipped through the whole rainbow sequence, from blue to red. Before 60,000 years, light actually weighed more than matter and dominated everything. Before 400,000 years, it was quite foggy, but at about that time, the fog lifted and space became transparent for the first time. To understand all these phenomena, we must visit the microworld of protons, electrons, and photons, for it is they that ultimately determine the sights and sounds of the newborn Universe.

Outline

I. In the last lecture, we learned how the microwave background directly reveals the million-year-old Universe. In this lecture, we'll visit that time and learn what it was like.

  A. There are three important differences between then and now.
   1. All the matter in a region of today's Universe was crammed in a region 1000 times smaller, with a billionth the volume.
   2. Unlike today, the matter was spread about very uniformly, a thin gas.
   3. It was several thousand degrees, everywhere.

  B. Here are the properties we'll be considering: composition, density, temperature, color, brightness, and transparency. In the next lecture we'll consider sound.

II. Let's start with the gas's composition.

    A. The atomic matter comprises 75% hydrogen and 25% helium, though it's so hot the atoms are ionized, just free electrons and nuclei.

    B. Although dark matter, neutrinos, and dark energy are all present, we wouldn't notice them, so they won't feature in this lecture.

    C. Photons are present in huge numbers. It's very bright!

III. Because many properties change with cosmic expansion, let's look at how to follow these changes.

    A. We use the scale factor, S. Recall, S was zero at the Big Bang, is one today, and tracks cosmic size. From Lecture Twelve, S increases rapidly at first and then more slowly. At 400,000 years, $S = 1/1000$. While S tracks cosmic size, S^3 tracks cosmic volume.

    B. Because density is mass/volume, it drops as $1/S^3$, as matter thins out. So we can figure out the gas density near 400,000 years.

        1. Since the volume was a billion times (1000^3) smaller than today, the density was a billion times larger.

        2. Today's average proton density is about 1 per 4 cubic meters. So a billion times larger is 250 protons per cubic centimeter (per sugar cube).

        3. This is very sparse by human standards: Our atmosphere has 10^{19} molecules per cubic centimeter.

        4. Only at 10 seconds was the proton density equal to our atmosphere's density.

IV. Now let's consider temperature.

    A. The young Universe was very hot, but it cooled as it expanded. At 400,000 years, it was 3000 K.

    B. If you sat in this gas, would it burn you? No, because it's too thin.

V. What about light?

    A. Light's brilliance would kill you instantly!

    B. The young Universe was bright because it was hot, and hot things glow.

C. From Lecture Four, a hot gas glows with a thermal spectrum: a broad range of colors, with a redder peak for a lower temperature gas.

D. During cosmic expansion, two things happen to the photons.
  1. They're redshifted.
  2. They get spread out.
  3. These combine to give a new thermal spectrum with lower temperature.

E. This explains why cosmic expansion cools the Universe.

F. Quantitatively, temperature is inversely proportional to scale factor: $T \propto 1/S$.

VI. What would we witness, floating in the young Universe?

  A. Light comes from every direction: a 360° panoramic bright sky.

  B. As the Universe cools, the sky color slips through a rainbow sequence, from blue to red, spending roughly 80,000 years in each color.

  C. The sky would be incredibly bright. At 100,000 years, the temperature is 6000 K, similar to the surface of the Sun. At that time:
    1. The sky appears as if covered by Suns—210,000 of them.
    2. It's also equivalent to diving inside the Sun's atmosphere.

  D. This awesome intensity of primordial light stems from expansion and red shift. Imagine halving S: Photon density increases by 8 (2^3), but with blue shift, photon energy density increases by 16 (2^4).
    1. So light's intensity changes with scale factor to the fourth power. This is a very sensitive dependence.
    2. For example, when $S = 1/10^3$, S^4 is $1/10^{12}$. The microwave background multiplied by 10^{12} is enough to kill you!

  E. With light so intense and the gas so thin, the actual ratio of photons to protons is about 1.6 billion to 1.
    1. This ratio doesn't change much as the Universe evolves, and it was set during the first second (for reasons we'll get to in Lecture Twenty-Nine).

    2. In our daytime atmosphere, there are 100 billion protons per photon—essentially the reverse to the gas in the young Universe.

F. Since both mass and energy make gravity, which is "denser": matter or light? Going backward in time, matter's density increases by $1/S^3$, while light's increases by $1/S^4$, so light's density increases faster.

    1. At 60,000 years, they're equal.

    2. At 10 seconds, when matter had our air's density, light was 200 times denser than lead!

    3. The transition between the radiation and matter eras leads to a particular quality that has been seen in the patterns of galaxies.

VII. Was the young Universe foggy or transparent?

  A. In a terrestrial fog, photons bounce off tiny water droplets. In the early Universe, they bounce off free electrons. So the gas was foggy.

  B. As the density dropped, one could see further.

    1. At 1 second, you can't see your hand in front of your face.

    2. At 100,000 years, you can see about 500 light-years.

  C. An important question is, can you see to the horizon? Recall, the horizon is how far light can travel since the Big Bang. So at 100,000 years you can only see 0.5% of the way to the horizon—so it's a relatively "thick" fog.

  D. Near 400,000 years, a dramatic event occurs: The free electrons are removed from the gas so the fog quickly clears. After that, the Universe is transparent.

VIII. This transition is called "recombination," and it marks the change from ionized to atomic, from foggy to transparent, from sound to silence, and from smoothness to roughness. It is an extremely important event.

  A. By 400,000 years, conditions are calm enough for electrons to be caught and held by protons, so atoms can finally form.

    1. Hydrogen and helium atoms are transparent to light, so from then on photons move uninterrupted in straight lines.

   2. After traveling for 14 billion years, affected only by red shift, they are the microwave background photons.

B. Consider our astronomer's view of the Universe:

   1. We're surrounded by a 3000 K wall of hot, glowing fog.

   2. In a static Universe, it would be like sitting in a 3000 K oven.

   3. Fortunately, red shift transforms this brilliant but lethal oven into a feeble microwave background.

Suggested Reading:

Hawley and Holcomb, *Foundations of Modern Cosmology*, chap. 12.

Mather and Boslough, *The Very First Light*.

Smoot and Davidson, *Wrinkles in Time*.

Hogan, *The Little Book of the Big Bang*, chap. 5.

Questions to Consider:

1. What was it like back in the million-year-old, infant Universe? Try to answer first in experiential terms—how it would have looked or felt—and then in more quantitative, physics terms—what were the temperatures, densities, colors, and so on?

2. Are you surprised at the degree of change between the infant Universe and today's Universe? Compare this to the difference between a newly fertilized egg and an adult human, or between the primitive Earth's surface and the current biosphere. In each of these, what are the primary reasons driving the enormous change? How similar or different are these kinds of processes? (Return to this question after you've finished the course, when you should understand more clearly what drives cosmic change.)

3. The brilliance of the 100,000-year-old Universe is like covering the sky with Suns. How many is that? Hint: The Sun is ½ degree across, so its radius is 15 arc minutes. How many square arc minutes is that (use πr^2)? To find out how many square arc minutes cover the sky, calculate how many square nautical miles cover the Earth's surface (use $4\pi R^2$, with $R = 3440$ nautical miles). Remember, 1 arc minute on the Earth's surface is 1 nautical mile, which is about 1.15 statute miles. You should find about 210,000 Suns cover the sky. The infant Universe is pretty intense!

Lecture Fourteen—Transcript
Conditions during the First Million Years

In the last lecture, we learned about our extraordinary ability to witness, directly, the preembryonic Universe, just 400,000 years after the Big Bang. By looking far away, we see the brilliant orange light from hot glowing gas, modified only by red shift of cosmic expansion, which stretches the orange light waves by a factor of 1000, so they arrive as microwaves. It's the famous cosmic microwave background. Although we concentrated on the observed properties of the microwave background, its spectrum and its panoramic image, we also began to get a feel for the conditions at that time. Most importantly, since it was well before any stars or galaxies had formed, it was radically different from today's Universe. Everywhere was simply an expanding, uniform, hot glowing gas.

If one wanted to look for parallels to traditional creation stories, some of which describe an initial era of chaos or void out of which the world formed, then here is such a parallel. Despite its utterly different form, emerging from this hot glowing gas over the eons came our current beautiful world, with its life-giving Sun and starry sky. Now, my aim in this lecture is to take you back to that early time, to try to experience what we might have seen or felt or even heard had we been alive back then. So let's start with a diagram that reminds us of the overall situation.

Imagine a huge region today, filled with galaxies. Here's the same region as it was in the early, million-year-old Universe. Now, there are three extremely important differences between these two regions. First, everything inside today's big cube was packed inside a tiny cube about 1000 times smaller. This diagram, in fact, is way off in scale. Now, look, here is the real difference: a 1-meter cube and a 1-millimeter cube—that's an amazing difference. That's a factor of a billion difference in volume. Now, second, the matter in today's cube is very clumped up—there are stars and galaxies separated by gaps. But in our smaller cube, the same matter is all smoothed out, completely uniform. And lastly, today's Universe is, on average, extremely cold, whereas back then, it was several thousand degrees everywhere.

So here are the properties we'll be looking at in the million-year-old Universe: composition, density, temperature, color, brightness, and transparency—and we'll save sound for the next lecture. So let's start with the gas composition. Here on Earth, the atmosphere you and I are now breathing contains about 80% nitrogen and 20% oxygen in the form of molecules. So what about the primordial gas? Well, remember, about 3 minutes after the Big Bang, a brief burst of nuclear fusion produced a composition of about 75% hydrogen and 25% helium, and not much else. Now, because of the intense heat, these gases were neither atomic nor molecular. They were fully ionized, just free electrons, protons, and helium nuclei, just zooming around.

Now, in addition to this, there was dark matter, outweighing the hydrogen and helium 6 to 1, but because it's so undetectable, apart from its gravity, I won't look at this much in this lecture. And the same is true for neutrinos—they're present in huge numbers, but we really wouldn't have noticed them because they don't interact with anything. The other dark component, dark energy, we'll also ignore in this lecture because back then, it had a much smaller density than all the other components. Finally, there are photons, light, in enormous numbers, and as we'll see, they are hugely important. So overall, we'll be dealing with a smooth gas of electrons, protons, helium nuclei, and photons, all zooming around, bashing into one another, effectively infinite in all directions.

Before we move onto the other properties, we need to get used to the fact that some of them will be changing as the Universe expands, and it will help if we have a simple way to follow these changes in our minds. Now, the best measure of expansion is our old friend S, the scale factor, or separation factor. It tracks how big the Universe is. As cosmic size doubles or triples, so S doubles or triples. And remember, S is 1.0 today, so during the first million years, S is roughly 0.001 or 1/1000. The Universe was 1000 times smaller than it is today. Now, this graph shows how S changes over time across the first million years. It's the curve we calculated back in Lecture Twelve, so it's 0 at the Big Bang, grows quickly at first, and then more slowly, and it passes 0.1% or 1/1000 near 400,000 years.

Now, if this curve, S, tracks the linear scale in the Universe, then S^3 tracks the volume. See, if we have a box and double its size, its volume goes up by 2^3, or 8. Triple its size and the volume goes up by 27. Now then, if that box contains gas, then the gas gets spread thinner throughout its increased volume, and so the gas density drops as the box gets bigger. It drops to 1/8 and then to 1/27, so it drops in proportion to the scale factor cubed. Now, this leads to a very handy formula: To find the cosmic density at any time, it's just the density today divided by the scale factor at that time cubed.

So with that background, let's ask a simple question: What was the density of the gas in the young Universe? Specifically, how many protons were there per cubic centimeter? That's how many protons in this cube of sugar—in the volume described by this cube of sugar, not the sugar itself. This is actually easy to figure out. Using the average density of protons in today's Universe, and that's 1 per 4 cubic meters, back then when the Universe was 1000 times smaller, those 4 cubic meters had 1000 times 1000 times 1000— or a billion—times smaller volume. So the density was a billion times larger, which works out to be about 250 protons per cubic centimeter.

So you can see the density dropping as the Universe expands on this graph, inversely proportional to S^3. It follows this orange line, which goes with the right-hand scale, from about 10,000 protons per cubic centimeter near 50,000 years to about 100 protons per cubic centimeter near 1 million years. Now, that actually sounds like quite a lot, but is it really? How many molecules do you think there are per cubic centimeter in our own atmosphere right here in front of me? Well, you may be surprised to hear that there are 10^{19}, a 1 followed by 19 zeros. That's a cubic kilometer's worth of sand grains, so cosmically speaking, we live in an exceedingly dense environment, and the primordial atmosphere is incredibly rarified by human standards. In fact, it's more rarified than the best vacuum that scientists can make on Earth. So does the primordial atmosphere ever equal our own atmospheric density? The answer is it does, but only 10 seconds after the Big Bang. So after only a few minutes, the Universe's density of protons and electrons is much lower than anything here on Earth.

Let's now consider the temperature during the first million years. Remember that the young Universe was very hot, but it cooled as it expanded, following this curve. At 50,000 years, it was 10,000 K. At 400,000 years, it was 3000 K. And by a million years, it was 1800 K. That's 1500°C, about the melting point of iron. Now, here's a question: Sitting in this gas at several thousand degrees, would it burn you? Surprisingly, the answer is "no," it wouldn't. It's far too thin. Remember, heat is associated with the jostling motion down in the atomic world, and the heat energy entering your skin, which determines whether you'd be burned or not, depends on two things—the average kinetic energy of each particle and the number of particles striking your skin per second. Now, in the early Universe, even though the kinetic energy of the particles is high because the gas is hot, there are so few of them hitting you that you'd barely even notice their heat at all.

However, despite surviving the hot gas, light would kill you instantly. So let's turn now to look at the primordial light. At various points in these lectures, I've commented that the young Universe was very bright. A "big flash" might be a better term than a Big Bang. Now, why was this? Why was there so much light created? The simple answer is that the Universe was born hot, and hot things glow, just like this piece of iron or this furnace. In Lecture Four, I mentioned why this happens. When two charged particles collide, like an electron and a proton, the collision creates a photon of light. Now, as long as the particles and photons interact and share their energy back and forth, they both arrive at what physicists call "thermodynamic equilibrium." Essentially, all particles have the same broad spectrum of energies and therefore have the same temperature.

Photons, for example, have a thermal spectrum like this, a broad range of colors with a peak. As you can see from these other examples, at lower temperatures the thermal spectra are weaker, and with longer—that's redder—peak wavelengths. Now we can understand why the Universe cools as it expands. Two things happen: The expansion causes all the photons to be redshifted to longer wavelengths, and they get more spread out—they are diluted. And together, these yield a weaker, redder spectrum, which is another thermal spectrum, but with a lower temperature. Bingo, cosmic expansion with its red shift and dilution of photons has just cooled the Universe.

Let's be slightly more quantitative. If we double the cosmic size, the wavelength of all photons also doubles, and so the wavelength of the peak of the thermal spectrum doubles—and this corresponds to half the original temperature. So we have a very important relation: The cosmic temperature is inversely proportional to S, the scale factor.

And it's this relation that gave our graph of cosmic temperature across the first million years.

Okay, so now, let's return to the simple question of "What would we witness in the young Universe?" Floating there, looking out into the distance, what would we see? Well, since light comes from every direction, you would see this extremely uniform glow all around you, rather like a 360° panoramic bright sky. Now, because the light has a thermal spectrum, this leads to a rather stunning spectacle. As the Universe cools, the whole spectrum shifts to the red, so the sky color changes. Here's what happens over the first million years. Starting at 50,000 years, the sky is blue. At 200,000 years, it's yellow. At 400,000 years, it's orange. And by a million years, it's crimson. The Universe's sky color slips through the rainbow sequence backwards, spending roughly 80,000 years in each color. This is a wonderful and beautiful quality of the young Universe. It actually paints the sky with a palette of human colors.

Now, although we'd recognize the color of the primordial sky, how bright would it be? I mean, could we stare at it without squinting? I'm afraid the answer is a very definite "no." Not only could you not stare at the sky, nor would it just blind you—it would incinerate you almost instantly. For example, around 100,000 years, the temperature was about 6000 K, which is similar to the Sun's surface temperature. So sunlight gives us a good idea of what 6000 K thermal emission looks like. So the primordial sky glows from every direction, with an intensity and color similar to the Sun. You see, it's as if you're standing on the Earth, not with just one Sun in the sky but 210,000 Suns—a sky filled with Suns, no gaps between them. This is what the young Universe is like. When I said it was fantastically bright, I really wasn't exaggerating.

Now, another way to experience this would be to dive down 1000 miles into the Sun's atmosphere. You'd be surrounded by a brilliant glowing gas. The light alone is powerful enough to evaporate you in 10 seconds. Here's a plot that shows all this. It uses the same line as the falling temperature, but it uses the right-hand scale to follow the sky brightness in noontime Sun equivalence. So going backwards in time, starting at 1 million years, one sees a panoramic featureless crimson sky as bright as 1000 Suns. By 0.5 million years, it's now deep orange and as bright as 10,000 Suns. At earlier and earlier times, the brightness soars upwards, while the sky color sweeps through the rainbow sequence. It's an utterly fabulous, if lethal, spectacle.

So why does the primordial light have such awesome intensity? As usual, it's due to expansion and red shift, so let's imagine going backwards in time, halving the scale factor, so volume gets smaller by a factor of 8. The number density of photons goes up by 8, but each photon's wavelength halves, so its energy goes up by a factor of 2. Combining these, the energy density of light increases by 16, which is a really big increase for just halving the cosmic size. We've actually come across this in Lecture Twelve already, when we saw that, unlike matter, whose density changes with the third power of the scale factor, the density of radiation changes with the fourth power. That extra power comes from red shift.

Now, the fourth power is the square of a square, which means that light's intensity is extremely sensitive to the scale factor. So going backwards to earlier times, light's intensity increases incredibly fast. For example, when the Universe was 1000 times smaller than today, the intensity of light was 1000^4—that's 1000 billion times higher than it is today. Well, if you take the microwave background and boost its intensity by 1000 billion, that's enough to kill you instantly.

So far, I've emphasized how brilliant the light is and how rarified the gas is, but let's be more specific. What's the relative number of photons to protons? What's the photon to proton ratio? Well, in today's Universe, it's just the ratio of the photon density in the microwave background, which is 410 per cubic centimeter, to the density of protons, which is 1 per 4 cubic meters. That makes a ratio of 1.6 billion photons for every proton. It's a huge number. Now, it turns out that that number doesn't change much as the Universe expands, so the photon to proton ratio is approximately fixed, and it's an important parameter for our Universe. And it was set, somehow, in the first second of the Big Bang.

Now, in Lecture Twenty-Nine, we'll explore why this number is so big, but for now let's just register the fact that photons are enormously numerous, more numerous than protons or electrons by about a factor of a billion. So imagine a cubic meter filled with sand. Let me take 60 grains of salt and throw them in. Those would be the protons compared to the photons. So cosmically speaking, matter particles are extremely rare compared to photons. Although that's true for the Universe as a whole, here on Earth, the situation is completely the opposite. We live in an enormously matter-dominated environment. Even under a noon-day Sun, our atmosphere contains 100 billion molecules of air for every solar photon.

Now, having stressed light's huge abundance in the Universe, it's important to ask whether it also contributes to the gravitational field. Remember that energy and mass are deep down the same thing, $E = mc^2$. And so any source of energy, when divided by c^2, gives a mass that in turn generates gravity. So a very important question is, "For the million-year-old Universe, which has more mass, matter or light?" As you may remember from earlier lectures, the answer to this question changes with time, as you can see in this figure from Lecture Four, which shows how the density of the major cosmic components changes as the Universe expands from left to right. Both axes are exponential, so it turns out that the densities follow straight lines.

So remember, then, dark energy is constant, so although it dominates today, it was irrelevant in the early Universe. Now, the other 2 lines are for radiation, which includes neutrinos, and matter—that's both atomic and dark. And both these increase rapidly, going towards earlier times, simply because the Universe was smaller. But notice that the line for radiation is steeper. That's because it's increasing with scale factor to the fourth power, while matter's density only increases with scale factor to the third power. So although radiation starts out today much less important than matter, the 2 lines cross when the Universe was 5000 times smaller than it is today. That's roughly 60,000 years after the Big Bang. And at that time, radiation and matter densities were equal, and of course, this separates the two great cosmological eras, the radiation era and the matter era.

Now, ultimately, what's happened is that at earlier times, the blue shift of each photon boosts its energy, and this gives light the advantage, and it guarantees a radiation-dominated early Universe. Here's one example I've used before. At about 10 seconds after the Big Bang, the proton density roughly matches our atmospheric density, but light was 200 times denser than lead. What an amazing situation. Now, you may worry that this is all too theoretical and impossible to verify, but that is not so. In Lecture Seventeen, we'll discover that this crucial transition near 60,000 years, when light yields its gravitational authority to matter, leaves its stamp on the pattern of galaxies across the sky. Essentially, it defines the boundary—about 200 million light-years—that separates small-scale inhomogeneity from large-scale homogeneity. But let's leave that topic to Lecture Seventeen and move on.

So the last property I want to talk about is whether the young Universe was foggy or transparent. Stated simply, could you see your hand in front of your face? Now, to answer that question, we need to know what causes fogginess. In the case of a terrestrial fog, photons deflect off tiny water droplets, so they don't travel very far in straight lines. They scatter around randomly like a pinball, defusing haphazardly through the fog. Now, what might cause photons to scatter in the early Universe, causing it to be foggy? Certainly, there were no water droplets. It turns out that free electrons scatter light. They are lightweight charges that wiggle when a photon passes by, causing it to be deflected. The process is called "Thomson scattering," after J. J. Thomson, who discovered the electron in 1897.

Back to the young Universe. With only a few thousand electrons per cubic centimeter, the fog is extremely thin by human standards. For example, after 100,000 years, the fog's visibility distance was about 500 light-years. Now, as the Universe expands and the density drops, this visibility distance increases, like the blue dashed line on this plot. At earlier times, the visibility distance gets much shorter. But only about 1 second after the Big Bang, could you literally not see your hand in front of your face? A rather moot example, since at that time, you'd evaporate in much less than a nanosecond.

Well, since there were no people then, arm's length is hardly a relevant distance, but what is relevant is the horizon distance. Remember from Lecture Eight that the distance light travels since the Big Bang is called the "horizon." It's as far as you can possibly see. At 100 years, it's roughly 100 light-years, the age of the Universe times the speed of light. So here's a question: Can you see to the horizon? If you can, then the Universe is effectively transparent. If not, then it's effectively foggy. Well, here is the horizon distance. It's basically the cosmic age in light-years, and it curves this funny way because the y-axis is exponential but the x-axis is linear. At 100,000 years, for example, the visibility distance of 500 light-years is less than 1% of the horizon distance. Now, that's equivalent to just 30 yards' visibility if you were standing on the seashore, so it is definitely foggy. But as time passes and the gas thins, you can see a larger and larger percentage to the horizon.

If we extend these lines, they'd cross near 45 million years, and you could finally see to the horizon. However, well before that, near 400,000 years, a dramatic event occurs, which makes the visibility distance do this. If you remove the water droplets, the fog clears. In this case, the free electrons are removed. They're caught by the protons to make atoms, and so the fog quickly clears. Cosmologists call this "recombination," and it marks an amazing transition from ionized to atomic, from foggy to transparent, from sound to silence, and from smoothness to roughness. It's one of the most important changes in the entire history of the Universe, so let's briefly look at what happens.

As you know, the early Universe was far too hot for atoms to form. Collisions by particles and photons just bashed them apart as soon as they formed. But as the Universe cooled, by about 400,000 years, the conditions were calm enough for electrons to be caught and held by the protons, and there was a frenzy of pairing up. The transition took about 80,000 years to complete, but after that the gas contained neutral atoms of hydrogen and helium, which is quite a different kind of gas.

Now, this atomic gas was completely transparent to the deep orange light that was present in the Universe at that time. So from then on, the photons were free to stream, without interruption, in straight lines as far as they could go. Almost all of them are still going, unchanged except by red shift, though a very few do end their long 14-billion-year journey rather abruptly when they run into your head as a microwave background photon. That, of course, is what the microwave background is. It's our direct view of that time. The light has come to us, uninterrupted, across a transparent Universe. The last time that photon was scattered was back in the 400,000-year-old Universe. In fact, cosmologists sometimes call the microwave background "the surface of last scattering."

Let me end this lecture by looking a little differently at this topic. In our astronomer's view of the Universe, we are surrounded by a 3000 K glowing gas. Now, in a static, nonexpanding Universe, its great distance wouldn't diminish its intensity. Remember from Lecture Eight, surface brightness is independent of distance. Naively then, we should see a sky covered by 3000 K Suns. We will be at the center of an utterly lethal oven, with 18 megawatts of power falling per square meter. Now, what saves us is cosmic expansion, which transforms this brilliance into an imperceptibly feeble microwave background.

Remember also from Lecture Eight, in an expanding Universe, the surface brightness decreases with the fourth power of the scale factor, so that when the oven walls are redshifted by a factor of 1000, the oven's brightness is reduced by a factor of 10^{12}. That oven of 18 megawatts per square meter becomes a refrigerator of 18 microwatts per square meter. It's the 3 K microwave background. So who would have thought that red shift alone would save us all from fiery incineration by the cauldron of creation?

Lecture Fifteen
Primordial Sound—Big Bang Acoustics

Scope:

This lecture looks at (and listens to!) the primordial sound waves oscillating in the hot glowing gas of the young Universe. After considering the nature of terrestrial sound, we're able to understand the nature of primordial sound. Although the Big Bang was silent, the roughness it introduced leads to sound. The hot gas falls into and bounces out of dense pockets of dark matter, making sound waves. These waves are visible on the microwave background as bright and dark patches, and analyzing them tells us about the sound. Its volume is rock-concert level, though its pitch is 50 octaves below the human range. Remarkably, it has a fundamental and harmonics, so it resembles a musical note! Computer simulations allow us to re-create the evolution of the sound from the Big Bang: It is a descending roar. Ultimately, the sound waves cease and turn into the first stars and galaxies.

Outline

I. In the last lecture, we learned the young Universe was filled with a thin, hot atmosphere. Were sound waves present? Yes, and they're the subject of this lecture.

   A. Cosmologists are interested in primordial sound because it can be used to learn about the properties of the Universe. That's the next lecture's subject.

   B. Primordial sound waves are also the precursors for the first stars and galaxies, which are the subject of Theme IV.

   C. In this lecture, we'll look at the nature and creation of primordial sound.

II. Let's begin with the nature of terrestrial sound.

   A. In our atmosphere, vibrating objects make sound waves.
   1. Peaks are where molecules bunch up, making higher pressure.
   2. Troughs are where molecules are further apart, making lower pressure.
   3. The pattern moves along a bit like a crowd surge.

    4. The speed of sound waves is similar to the speed of molecules.

B. Some more terminology.
    1. The distance between two peaks is the wavelength (λ).
    2. The number of waves that pass per second is the frequency (f).
    3. Wavelength and frequency are complementary: When one is big the other is small.

C. Three important properties of sound are loudness, pitch, and quality.
    1. Loudness depends on the ratio of pressure variations (peak-to-trough) to the average pressure. It is measured in decibels.
    2. Pitch depends on frequency, which is measured in hertz.
    3. Quality depends to the particular set of frequencies present. The same note on a piano and violin has different "quality" because the higher frequencies are stronger in the violin.

D. A good way to analyze sounds is to use a "sound spectrum." All sounds have many frequencies present simultaneously, and the sound spectrum shows how loudness (y-axis) depends on frequency (x-axis).

E. Musical notes have sound spectra with a set of equally spaced, almost pure tones: a fundamental and harmonics. This is typical of resonant systems, which have just a few ways of vibrating.

F. Noise, like wind in the trees, has a broad, smooth sound spectrum with little or no structure.

G. As we'll see, primordial sound has a quality somewhere between a musical note and noise.

III. Primordial sound.

A. The young Universe was filled with a hot, thin glowing gas in which sound (pressure) waves could move.

B. Because the gas was foggy, protons, electrons, and photons were all tied together in a single "photon-baryon" gas.

C. Because pressure depends on particle density and temperature ($P \propto n \times T$), and because photons vastly outnumber protons or electrons, then it is *light* that dominates the pressure. The pressure was low by human standards: just a millionth of our atmosphere.

D. Because the speed of pressure waves is roughly the particle speed (photons in this case), the sound speed was very fast: about 60% of light speed.

E. In sum, we have a low-density, foggy gas of light and matter, all behaving as a single fluid with modest pressure and very high sound speed.

F. We see the peaks and troughs of primordial sound waves as bright and dim patches on the microwave background. They are caught as they cross the wall of fog, made visible because the gas is glowing.

IV. Primordial sound loudness and pitch.

    A. Loudness is measured from the variation in brightness of the patches on the CMB: about 1/10,000. This corresponds to 110 decibels: rock-concert loudness!

    B. The patches reveal wavelengths from 40,000 to 700,000 light-years, with oscillation periods of about 100,000 years: 50 octaves below human hearing!

    C. The sound is so deep because the Universe is so big. Its "organ pipes" are roughly horizon-sized.

V. Generating primordial sound.

    A. The Big Bang is a misnomer: The initial expansion of space was silent.

    B. However, the Universe was born slightly lumpy: a 3-D mottle of denser and less-dense regions.

    C. The photon-baryon gas falls into, and bounces out of, the denser pockets of dark matter.

    D. The many dark matter pockets of different size act like a register of organ pipes, each with its own note.

VI. The primordial sound spectrum.

    A. Analyzing the CMB patchiness yields a sound spectrum, $C(\lambda)$, the strength of waves of different wavelength.

        1. $C(\lambda)$ looks like a cross between a musical note and noise. It has a fundamental and harmonics that are all broad.

        2. We can re-create the sound by shifting up 50 octaves. It sounds like a deep roar.

    B. The most remarkable features are the harmonics. What makes these?

        1. Vibrating systems have harmonics because they are bounded in space, allowing 1, 2, 3, and so on waves to fit between the boundary.

        2. Although the Universe has no *spatial* boundary, it is bounded in *time*. At a given time (e.g., the CMB), regions of a specific size are caught at maximum or minimum compression or rarefaction. These specific region sizes form the harmonics.

    C. Despite the harmonics, the sound isn't musical for several reasons.

        1. Other, nonacoustic, effects enter the CMB patchiness affecting $C(\lambda)$ and broadening the harmonics.

        2. The pure sound, $P(k)$, can be generated using computer simulations of the young Universe, but it is still noiselike.

        3. The main problem is that the harmonics are broad, and our ears hear the range of frequencies as noise.

        4. For fun, one can narrow the harmonics to make a "musical" version of the sound. It has a rather eerie quality.

VII. Although we only witness the sound waves at 400,000 years on the CMB, we can use the computer simulations to follow the changing sound, starting at the Big Bang.

    A. The overall impression is of descending pitch.

    B. This arises because sound waves can cross and activate larger regions. As time passes, larger organ pipes are added to the register.

C. The drop in pitch slows down because the sound speed slows, and so the growth of active regions also slows.

D. At 400,000 years, when the fog clears, the sound oscillations cease and ultimately turn into full-scale collapse to make the first stars and galaxies.

E. We end the lecture with more fun by replaying the evolving sound with narrow harmonics, first including the downward slide and then removing it to hear the subtly changing chord.

Suggested Reading:

Hawley and Holcomb, *Foundations of Modern Cosmology*, chap. 14.

Hu and White, "The Cosmic Symphony."

Questions to Consider:

1. We make sound by using our diaphragm muscle to push air past our vocal chords, making them vibrate. What makes primordial sound? Where does the energy come from to drive the sound waves, and why do they oscillate?

2. Why is the Universe initially silent and only slowly begins to make sound? Why does the evolving sound drop in pitch, with deeper notes added as time passes? Why does the sound quickly cease near 400,000 years?

3. Let's figure out the pressure in the primordial gas near 100,000 years. The pressure in any gas is proportional to $n \times T$, where n is the number of particles per cubic centimeter, and T is the temperature in Kelvin. At 100,000 years, we have $T = 6000$ K and $n = 5 \times 10^{12}$ photons per cm^3. Remember, Earth's atmosphere contains 10^{19} molecules per cm^3 at a temperature of about 300 K. Use these values to figure out the primordial pressure in atmospheres. You should get about 10^{-5} atmospheres. Since 1 atmosphere is 23,000 pounds per m^2, what force does the primordial gas press on each square meter? Could this break a 1-square-meter sheet of tissue paper?

Lecture Fifteen—Transcript
Primordial Sound—Big Bang Acoustics

In the last lecture, we took a trip to visit the million-year-old Universe, to take in the sights, at least in our mind's eye. We found a slowly changing rainbow sky; an exceedingly thin, slightly foggy hot atmosphere; and everywhere brilliant light, as bright as hovering over a star's surface. Here, within the Earth's atmosphere, we're constantly aware of sound waves coming from all directions. Were there sound waves in the primordial atmosphere of the young Universe? The answer is "yes," there were, and they're going to be the subject of this lecture and the next.

Now, quite apart from the delightful prospect of actually listening to primordial sound, there are a number of reasons cosmologists are interested in Big Bang acoustics, and the primary one is that the sound an object makes tells you a great deal about its nature. So by studying primordial sound, you can actually measure many important properties of the Universe, and we'll look at how this is done in the next lecture. But in this lecture, we're going to look at the nature of the primordial sound itself and how it's generated.

Now, sound waves inherently involve slight variations from place to place, so primordial sound also provides our first indication that the Universe isn't completely smooth. And as we'll learn in Lecture Seventeen, over time the sound waves grow to form the first stars and galaxies, so they act as a crucial bridge from the smooth, young Universe to the lumpy, adolescent Universe. These next two lectures will also set the stage for Theme IV, which is all about the formation and evolution of galaxies and the galaxy web.

So let's begin with the nature of sound. This little animation shows sound waves coming from a tuning fork. The wave peaks are where molecules bunch up in compressions, giving a slightly higher pressure. The wave troughs are where molecules are further apart in a rarefaction, giving a slightly lower pressure. The pattern moves along a bit like a crowd surge: Molecules bash their neighbors, who bash their neighbors, and so on. As you might imagine from this crowd surge metaphor, the speed of a sound wave is similar to the speed of the individual molecules. For example, the speed of sound in air is about 750 miles per hour, which is similar to the 1000 miles per hour of a typical oxygen or nitrogen molecule.

Here's a bit more terminology. The distance between two peaks is called the "wavelength," and the number of waves that pass by per second is called the "frequency." It's measured in hertz, which are really waves per second. Wavelength and frequency are complementary, so big waves have small frequencies, and small waves have big frequencies. Now, there are three important properties that sound has that we need to consider:

First is loudness, which depends on the ratio of pressure variations—that's peak to trough—compared to the average pressure, which I've written as $\Delta P/P$. It's measured in decibels. Going from deafening to noisy to quiet is 135 to 75 to 15 decibels, and these have pressure variations of about a thousandth, a millionth, and a billionth of an atmospheric pressure. So these are tiny variations. Human ears are very sensitive and can cope with a huge range of pressure differences.

Next is pitch, which depends on frequency—the number of waves that pass by your ears per second. That's about 200 waves per second. That's about 600 waves per second. Concert A, for example, is 440 hertz, while 220 hertz is an octave below, and 880 hertz is an octave above. Normally, humans can hear sounds with a pitch somewhere between about 20 hertz and 20,000 hertz (or 20 kilohertz).

The third property of sound is its quality. Why does A440 played on a piano sound different from A440 played on a violin? Each has a strong 440-hertz component, but many other frequencies are also present. In fact, almost all sounds have many frequencies present simultaneously, and the best way to show them all is using a sound spectrum.

Now, sound spectra are going to be important in these lectures, so let's take a moment to get familiar with them. Here's the sound spectrum of a flute playing a single note, and here's the note itself. Along the bottom is frequency, with low frequencies, low pitch, on the left and high frequencies, high pitch, on the right. The numbers here are kilohertz, or thousands of hertz. And up the axis is loudness in decibels. So all sounds have a spectrum, with many pitches present, some louder than others. Now, in the case of this musical note, there is a very clear pattern made from a set of almost pure tones. There's a strong fundamental, which our ears hear as the primary pitch, and a set of higher harmonics.

This is, in fact, typical of many resonant systems. They have a number of specific waves of vibrating, each with its own frequency. Resonant systems are things like stretched strings and columns of air in musical instruments, or objects with well-defined boundaries, like a wine glass or a drum skin.

Let's now play the same note, but on a clarinet. The subtle difference in sound is because the relative strengths of these harmonics are different. The difference in turn depends on the different structure of the flute and clarinet and would be different again for a bassoon or a piano.

The other, more general, category of sound is noise—wind in the trees, or a waterfall. These sounds have a wide, continuous range of pitches, and here's an example of a broad, smooth sound spectrum, with little or no structure, and here's its sound.

Now, in a minute, we'll learn that the primordial sound has a quality somewhere between a musical note and noise, but first let's return to the young Universe and try to see how sound waves can be present. Now, first of all, you must please forget what you may have learned as a child, that there is no sound in the vacuum of outer space. In the last lecture, we learned that in the young Universe, all matter was spread out evenly, in a sort of atmosphere. It was extremely thin by human standards, but nevertheless, sound waves could move in it. Let's be a bit more specific.

At 100,000 years after the Big Bang, each cubic centimeter contained about 2000 protons and electrons and about 3000 billion photons, all at 6000 K. It's a hot, thin, brilliant foggy gas. Now, the fact that the gas is foggy means that all three components are tied together. All particles are bouncing off each other and behave as a single gas. Cosmologists call this a "photon-baryon gas." Baryon, you may remember, is just a fancy word for the atomic constituents: protons, neutrons, and electrons. You might think of the multicomponent gas a bit like our own atmosphere, where the oxygen and nitrogen molecules are intimately tied together as a single gas.

Now, if we're interested in sound, we need to ask what the pressure of the gas is. Pressure is the force that particles make when they hit a surface, like rain striking a windowpane. Each drop pushes on the window. Ultimately, it's this force that makes the gas move back and forth, from peak to trough in a sound wave. Now, it turns out that the pressure of a gas depends only on two things: its temperature and how many particles there are per cubic centimeter. And perhaps surprisingly, for any given temperature, it doesn't matter what kind of particle—it could be a proton, an electron, or even a photon. They all contribute the same pressure, and this leads to some rather remarkable facts about the primordial gas.

First, because photons of light far outnumber all the other particles, they also provide almost all of the pressure, and this seems a little unusual or strange to us. We don't normally think of light beams— like flashlights—as things that push, but they do, just a tiny bit. And when the light is brilliant and the atoms are rare, then light can dominate the pressure, as it does in the young Universe. Secondly, the actual pressure in the primordial gas is very low by human standards, about a millionth of our atmosphere's pressure. And thirdly, since it's light that provides the pressure, the speed of the pressure waves is incredibly fast. Remember our crowd surge metaphor? Pressure waves move with roughly the speed of the particles, and in this case, the particles are photons, moving at light speed. So the speed of sound waves in the primordial gas is about 60% of the speed of light.

So summarizing, we have an extremely lightweight, foggy gas, a brilliant light, and a trace of particles—all behaving as a single fluid, with a modest pressure and very high sound speed. With light dominating, you can also think of the primordial sound waves as great surges in the brilliance of light. Now, with this background now in place, how do we observe the sound waves in the young Universe?

Well, here's our cleaned, stretched image of the microwave background, showing the famous patchiness. You can think of the patches as the sum of many waves, like these water waves, but instead of water height, the waves are of higher and lower pressure. These are sound waves, caught and made visible as they cross the wall of fog that is the microwave background. The reason we see them as brighter and darker microwave emission is that regions of higher and lower pressure also have higher and lower temperature, which in turn makes the gas glow brighter and dimmer. So the patches on the microwave background are the peaks and troughs of sound waves, made visible because the gas is glowing.

So let's now figure out what the primordial sound actually sounded like. First of all, finding its loudness is easy. The variation in brightness, which is roughly 1 part in 10,000, tells us more or less directly what the variation in pressure is, $\Delta P/P$. Even though these variations are very slight, they correspond to about 110 decibels. That's about as loud as a rock concert, front-row seats at Pink Floyd. It seems delightful to me that the primordial sound is neither disappointingly quiet and nor is it fatally loud—it's just viscerally powerful.

Now, finding pitch is actually also pretty easy, because we know the distance to the microwave background, and so we can convert the angular size of the patches on the sky to their physical size in light-years. So we have their wavelengths. Now, typical primordial sound waves are between 40,000 and 700,000 light-years long, so these are huge waves, and since we also know the speed of sound— that's 60% of the speed of light—we can figure out their frequency.

So a typical sound wave may take 100,000 years to pass you by, so the primordial sound is way too deep for us to hear. Not even a single wave would pass by you during your entire lifetime. There's actually a simple reason the primordial sound is so deep—the Universe is so big. Just as larger air pipes make deeper notes, the Universe's organ pipes are roughly the horizon size—huge. So not surprisingly, they make very deep notes, roughly 50 octaves below our normal range of hearing.

So let's now look at how primordial sound is generated and then sustained. Now, you may think this is a bit of a nonquestion. "Well, there was a Big Bang, and the explosion must surely have been very loud." Well, this is completely wrong thinking. The Big Bang was not an explosion into an atmosphere. It was an expansion of space itself. In fact, the Hubble law tells us that every point recedes from every other point—no compression, no sound. Paradoxically, the Big Bang was totally silent. So how does the sound get started? Well, in Lecture Thirty-Two, we'll learn that although the Universe is born silent, it was also born very slightly lumpy. On all scales, from tiny to huge, there were slight variations in the density of all components: randomly scattered everywhere, a sort of 3-dimensional mottle of denser and less-dense regions.

But, and this is important, the uneven dark matter and uneven photon-baryon gas, behave quite differently. The dark matter clumps basically stay in place, while the photon-baryon gas, because it has pressure, bounces in and out of the dark matter clumps, making sound waves. This shows a bit more detail. Here's a region where the dark matter is a little denser. Now, the gas feels the dark matter's gravity, so it begins to fall towards this denser region. Soon, though, its pressure is higher, and this halts and reverses the motion, pushing the gas back out. Now, this time, it overshoots, only to turn around and fall back in again. The cycle repeats, and we have a sound wave. You can think of this denser region a bit like a spherical organ pipe, into which gas falls and bounces in and out.

Now, of course, there isn't just a single denser region—there's a whole range of dark matter clumps, of all sizes and all shapes. And so in a sense, there is an entire register of organ pipe sizes. We seem to have an entire musical instrument here, with a wide range of possible sounding notes. So this brings us to our next important question: What is the primordial sound spectrum? Which of these organ pipes is sounding, and which are not?

So let's start with the microwave background patchiness. You can think of this image as being made from the sum of a few very long waves of low frequency, many medium waves of medium frequency, and many small waves of high frequency, all added together. Each sized wave contributes a particular part of the sound spectrum. Now, using a clever computer program, it's actually possible to disentangle all these waves and see how much there is of each size, how loud each pitch is. The resulting sound spectrum is called $C(\lambda)$, where C is the loudness and λ is the frequency. Basically, big patches on the microwave sky have a small λ, and small patches on the microwave sky have big λ.

And here's the amazing result, which includes the best data, as of about 2006. And the line is a theoretical computer model, which we'll talk about later. It looks like a cross between a noise spectrum and a musical instrument spectrum. It has an overall broad shape, but with an unmistakable fundamental and harmonics. This has got to be one of the most stunning results for modern cosmology. It's the birth cry from the young Universe.

Well, to make this audible, we need to add all these tones together and shift up by about 50 octaves. And to do this, I've simply made the λ value of patch frequency into hertz, so the first peak falls close to 220 hertz. That's the A below concert A. Let's play that sound now, and remember—to be authentic, it needs to be at rock-concert volume, so here it is. Well, I'm not sure what you made of that, but it's certainly more noise-like than musical, a sort of deep roar. But perhaps the most bizarre feature of primordial sound is those harmonics—specific tones, roughly equally spaced in pitch—that stand out. What makes those?

Now, for musical instruments, harmonics arise because the vibrating system is bounded in space, allowing exactly 1 or 2 or 3 or any whole number of waves to fit within the boundary. But the Universe has no boundary, so why the harmonics? It's because the Universe is bounded in time. Between the Big Bang and the microwave background is 400,000 years. During that time, the gas is busy, falling into and bouncing out of a wide range of dark matter organ pipes. For the smaller ones, there's been time for several oscillations, while for the larger ones, there's only been 1 or 2. Now, there is the largest region into which the gas has just fallen for the first time. These regions have maximum pressure contrast, and they show up as the fundamental, about 700,000 light-years in size, 1° on the sky.

There's also a somewhat smaller region size, where the gas has fallen in once, bounced out, and fallen in for the second time. These sized regions show up as the third peak. Now, likewise, there's a whole sequence of ever-smaller regions, where the gas is just arriving at the bottom for the third, fourth, and fifth time; and these yield the fifth, seventh, and ninth harmonic peaks. The odd harmonics are called "compression peaks" because the gas in the dark matter regions is compressed. Now, there's a whole set of intermediate-sized dark matter clumps, where the gas has just bounced out at the time of the microwave background, leaving a rarefaction, an under-pressured region. In fact, the second, fourth, sixth, and eighth harmonics arise this way. So the even harmonics are called "rarefaction peaks," because the gas in the dark matter regions is in the rarefaction part of its cycle.

So ultimately, dark matter clumps with just the right size to be caught at the time of the microwave background, at either a maximum compression or a maximum rarefaction, exhibit the strongest patchiness, and so these sized regions appear as peaks in the sound spectrum. It's perhaps disappointing that the sound isn't more musical to our ears, especially since it contains these harmonics. So let's look at why that's the case.

First of all, the microwave background patches include several contaminating effects, and they only approximate the true sound. One way to access the true, or pure, sound uses a computer to calculate the sound in the young Universe. Now, we'll talk more about these computer simulations in the next lecture, but here you can see the sound spectrum of this pure sound, which cosmologists call $P(k)$. Well, does the cleaner set of harmonics seen in the pure sound, $P(k)$, make a more musical sound? Let's see by alternating between the two. Well, to my ear, the pure sound is a little cleaner, but it's still far from musical.

The main problem is that the primordial harmonics are broad, and our ears hear this range of frequencies as noise. This is actually very obvious if we use a decibel scale to compare the primordial and the flute spectra. The flute harmonics are so much narrower. Listen to the difference. Now, at this point, we can actually have a little bit of fun and artificially make the primordial harmonics narrow to see what a musical version of the sound would be like. Keeping their pitch and relative intensity the same, here it is. It has a rather eerie quality, containing an interval between a musical major third and minor third, something one rarely comes across in Western music.

Okay, what about the primordial sound at earlier times, before we see it on the microwave background? You see, we can't actually access this sound observationally because it would need us to look deeper into the fog, but what we can do is use those computer simulations to re-create the sound. Now, in this next example, those 400,000 years are compressed into just 10 seconds. And to go with the sound, there's a movie of the changing sound spectrum. The movie also shows you the changing sky color and sound waveform. So here it is. Well, let's do that again, since it passes rather quickly. This time, you might also follow the little trackers for time and scale factor at the top left.

Well, the primary impression is of a descending pitch, so what causes this? The effect is actually similar to first playing a violin, switching to a viola, and then a cello, and then a bass. The bigger the instrument, the deeper the sound. The dark matter clumps into which the gas is falling become larger at later times. What's happening is that as time passes, a sound wave can cross a larger and larger region and so activate larger and larger dark matter regions. In fact, if you think of the Universe as containing an entire set of organ pipes of all sizes, from tiny to huge, only those smaller than the sound horizon—that's the distance that sound can travel since the Big Bang—are actively sounding. As the sound horizon grows larger and larger, larger and larger organ pipes are added to the register, and so the sound pitch drops.

You may have noticed a more subtle effect: The drop in pitch slows down and stabilizes. This is because the sound speed is dropping, and slowing the growth of the sound horizon. Now, the reason the sound speed is dropping is that atomic matter begins to weigh down the gas as light's power begins to wane. Ultimately, as the fog clears and the light can no longer push on the gas, the sound speed plummets, and the sound horizon freezes, and the pitch also freezes. Ultimately, as we'll come to in Lecture Eighteen, the oscillations cease and turn into a full-scale collapse to form stars and galaxies. So these evolving sounds are a necessary precursor to stars and galaxies.

Let's end with a little more recreational fun and games by rendering these evolving sounds in musical form, with the harmonics artificially narrowed. Now, the frequency, or pitch, of each harmonic is plotted vertically, with hertz on the left and the piano keyboard on the right. The time axis is exponential, so we have about 2 seconds for each factor of 10 in time. First is a version that includes the drop in pitch. And now the same version, but with the fundamental held fixed at 220 hertz to help us hear the changing chord. Finally, let's force these frequencies onto the notes of our musical scale, which introduces a slightly artificial sense of rhythm as they shift to whichever note is closest. Now, this, I'm afraid, is as close as it gets to something even remotely musical.

Well, let's quickly summarize, using these figures. The WMAP satellite creates an image of the microwave sky, which after suitable cleaning and stretching reveals a myriad of brighter and darker patches, like rough water on an ocean. These are, in fact, the peaks and troughs of sound waves, caught as they cross the wall of glowing fog that is the microwave background. An analysis of the patterns of patches reveals a broad sound spectrum, with a fundamental and harmonics. The pressure variations correspond to about 110 decibels, rock-concert loudness, and the wave frequencies are 50 octaves below human hearing. Nevertheless, it's possible to upshift in pitch and play this sound. It's rather rough, raw. One can use computer simulations to access the pure sound and even generate its drop in pitch across the first half-million years, but the broad harmonics always sound rough to our ears.

In the next lecture, we'll learn that slight differences in the properties of the Universe lead to slightly different sound spectra, and so we'll see how cosmologists use the sound spectra to actually measure the cosmic properties. Now, finally, however you reacted to hearing these versions of the primordial sound—amused or impressed or perhaps slightly disappointed—as we'll learn in later lectures, it had to be exactly that way if the Universe was going to become the huge and stunningly creative place that it is.

Lecture Sixteen
Using Sound as Cosmic Diagnostic

Scope:

For cosmologists, the primordial sound spectrum provides a gold mine of information. It is not difficult to see why: The sound an object makes—whether the Universe or a wine glass—depends on its properties. If you understand how sound is made, you can use it to figure out the properties of the vibrating object. This method has been used with stunning success in cosmology. A crucial ingredient is an accurate computer simulation of the early Universe, and thanks to several decades of hard work, such simulations now exist. We'll look at how these are run and the kinds of effects they must consider. Ultimately, by varying the computer Universe's sound to match the observed sound, one can measure many fundamental cosmic properties. The ability of the simulation to match the data is so good it should convince you that our story of the young Universe is largely correct.

Outline

I. Cosmologists are interested in primordial sound mainly because it can be used to figure out the properties of the Universe.

    A. Strike a wineglass and a teacup—their different sounds betray their different properties. The same is true for the Universe: If its properties were different, its sound and sound spectrum would be different.

    B. Our aim is to reverse the process, to use the sound spectrum to figure out the Universe's properties.

    C. There are three crucial steps.

        1. Understand the physics of the young Universe.

        2. Insert this physics into a computer simulation that can calculate the sound spectrum.

        3. Adjust the simulation so that the calculated sound spectrum matches the observed sound spectrum. That simulation then reveals the Universe's true properties.

II. Rather than delve into the real physics, here's a metaphor. One can consider the microwave background as a primordial document, written by Nature in its own language: physics. To read this document one must be "fluent" in physics.

    A. Fortunately, much of the relevant physics is surprisingly simple.

        1. The young Universe is extremely smooth.

        2. There are really only two components: the photon-baryon gas and dark matter.

        3. The gas is in thermal equilibrium.

        4. Deviations from exact smoothness are very slight.

    B. Despite these simplifying qualities, it still took 30 years for theoreticians to construct a detailed treatment of the hot expanding gas laced with slight fluctuations.

III. Using this knowledge of physics, one must calculate the primordial sound spectrum and its appearance on the microwave background. It took 15 years to develop the computer programs to do this, and the most famous one is called CMBFAST. It is, in fact, publicly available, so you could run it yourself! There are three stages.

    A. One must specify the parameters of the Universe to be simulated; things like H_0, the various omegas, and an input roughness injected by the Big Bang.

    B. The program then calculates how the roughness evolves over time, generating what in the last lecture we called $P(k)$: the true sound spectrum.

    C. Finally, these sound waves are projected onto the microwave background to reproduce what we actually see. More technically, we go from $P(k)$ to $C(\lambda)$.

IV. This last step is very important, so let's look at it further.

    A. A metaphor helps. Imagine a concert hall with bad acoustics and a symphony playing on the stage. The sound we hear, as we sit at the back, is distorted. Similarly, the sound we measure from the microwave background, $C(\lambda)$, is a distorted version of true sound, $P(k)$.

B. If you already know what the symphony should sound like, then the distortions help to reveal the nature of the concert hall. Similarly, we can learn even more about the Universe from these distorting effects.

C. CMBFAST includes 8 or 9 such distortions, but we'll look at just 4.

D. The first is pretty straightforward: Within a sound wave, gas moves, and this motion causes Doppler shifts.

    1. These Doppler shifts add their own patches to the CMB.

    2. In the sound spectrum, they cause $C(\lambda)$ to have less prominent harmonics than $P(k)$.

E. The second is called the "Sachs-Wolfe effect."

    1. Photons leaving denser regions lose energy as they struggle against the pull of gravity.

    2. This adds patchiness to the CMB: Denser regions appear dimmer.

    3. Most of the patchiness you see in the all-sky CMB images is due to this effect. The actual sound waves are all smaller than 2°.

F. The highest-frequency sound waves are suppressed, for two reasons.

    1. The wall of fog is many thousands of light-years thick, so sound waves smaller than this get lost within the wall and appear washed out.

    2. After 400 million years, the first stars reionize the gas, and this introduces a thin mist in front of the CMB. This tends to blur the finest details, which include the smallest waves.

G. These effects (and others) are well understood, so CMBFAST is able to calculate the sound spectrum seen in the CMB from the true sound spectrum it calculated for the gas itself.

V. The final step is to compare the CMBFAST-calculated sound spectrum with the actual observed sound spectrum from the CMB. The simplest way to show this is to vary just one parameter at once and watch the change in sound spectrum. Here are three examples.

A. Total cosmic density.

    1. As we increase total cosmic density, the sound spectrum shifts to lower frequencies: Denser Universes have deeper voices.

    2. Matching to the observed data singles out a total density to 1% accuracy.

    3. Listening to acoustic versions: The change in pitch is obvious.

B. The fraction of atomic matter.

    1. As we increase the atomic fraction, the fundamental gets much stronger and the location of the higher harmonics changes.

    2. Matching to the observed data reveals an atomic content near 4%.

C. The time at which the first stars form.

    1. Recall that the first stars reionize the gas and that this creates a foreground mist through which we see the CMB. The earlier this occurs, the thicker the mist.

    2. Increasing the thickness makes the CMB patches more blurred and suppresses the higher frequencies.

    3. Matching to the data suggests the mist is 92% transparent, so the first stars formed around 400 million years, with about 30% uncertainty.

VI. Let's return to our graph, showing the best microwave background data and the best model fit as of about 2007.

A. The model has about 8 parameters, many of which are determined to a few percent accuracy. I'm postponing to Lecture Twenty-Six a full discussion of these. But let's step back and grasp the bigger picture.

B. First, the data. It's amazingly good, with tiny error bars over much of the spectrum. Primordial sound is now well measured, with its harmonics, low-frequency plateau, and high-frequency damping.

C. But more impressive is the calculated spectrum. It fits the data perfectly, matching all the details. This, more than anything, should convince you we're on the right track. Our story of the first million years, with its brilliant fog and surging sound waves, is not just a modern myth. It all actually happened!

Suggested Reading:

Hawley and Holcomb, *Foundations of Modern Cosmology*, chap. 14.

Silk, *The Infinite Cosmos*, chap. 13.

Jones and Lambourne, *An Introduction to Galaxies and Cosmology*, chap. 7.

Questions to Consider:

1. How do cosmologists use the patchiness on the microwave background to measure many of the Universe's properties? Although one might summarize the process as "the sound the Universe makes tells us about its properties," in practice a great deal is involved. What basic things need to be done to go from the patchiness on the microwave background to a list of measured cosmic parameters?

2. The microwave background gives us a rather corrupted view of the primordial sound waves. What are some of the processes that play a role in this corruption, and how can they be used to learn even more about the properties of the Universe?

3. Why are the patches on the microwave background larger than about 2° not, in fact, sound waves? What makes these larger patches?

4. Make your own Universes! Go to the Lambda website (http://lambda.gsfc.nasa.gov/toolbox/tb_cmbfast_form.cfm) and run your own CMBFAST model Universe. First use all the default parameters, and click "run" at the bottom. This specifies the concordance model, and the output gives the famous graph with fundamental and harmonic peaks. Now try varying some of the parameters—for example, make all the matter atomic, or set omega-vacuum to zero. Notice how the sound spectrum changes, and see if you can reproduce some of the trends discussed in the lecture.

Lecture Sixteen—Transcript
Using Sound as Cosmic Diagnostic

This is the last of our four lectures on the microwave background in the first million years. To be honest, so far we've taken a bit of a tourist route through the subject, taking an imaginary trip back to those early times to see the sights and hear the sounds. Now, while that's a reasonable approach for a lecture course like this, you won't be surprised to hear that it isn't the path most cosmologists spend much time walking down. In the professional community, the microwave background has one outstanding quality: It contains a huge amount of information about the Universe. And in recent years, Herculean effort has been put into accessing that information, to use the microwave background, to actually measure many fundamental cosmic properties. It's this more pragmatic theme that I want to explore in this lecture—how the properties of the microwave background tell us about the properties of the Universe.

Now, let me start by illustrating the main aim of microwave background studies. Here is a wineglass and a teacup. Let's make each vibrate. Let's make it sing in its own voice. Well, of course, they sound different because they have different structures and different compositions. The material is different. This is true for all objects. They each have their own unique sound and sound spectrum. Now, the same is true for the Universe. If the Universe was slightly different, let's say it had more atomic matter or more dark energy, then it would sound different. We would see a different pattern of patches on the microwave background, which would have a different sound spectrum. So the whole approach is to turn all this around and try and work backwards, using the observed sound spectrum to figure out the properties of the Universe. For our wineglass and teacup, if you were an expert on vibrating structures, you could listen to those sounds, or perhaps study their sound spectra, and figure out the structure and composition of each.

So can we use this kind of approach to figure out the properties of the Universe? Well, as you'll see, the answer is a very definite "yes," and the method lies at the heart of Big Bang acoustics. Now, this is where our tourist approach really falls short. In the last lecture, when I showed you the sound spectrum, and you heard the sound, you didn't immediately go, "Ah yes, 4% atomic matter, 23% dark matter." So it doesn't work. I'm afraid a simple intuitive approach just isn't appropriate. In fact, there are three crucial steps to the professional approach, and here they are:

- You must first thoroughly understand the relevant physics, in particular, how sounds are generated and changed during expansion.
- You must then put all that physics into a computer simulation, which reproduces, as closely and accurately as possible, the sound in the young Universe.
- And finally, you must adjust your simulation so that its calculated sound spectrum matches, as closely as possible, the observed sound spectrum. When the match is perfect, then hopefully these model properties are also the properties of the real Universe.

So for the rest of this lecture, we're going to look at each of these three steps in more detail. Let's start with the first, somewhat daunting step of understanding the physics, but I'm going to recast it in a slightly novel way that bypasses all the details, using a linguistic metaphor. So you want to think of all the photons that rain down on us from the distant Universe. The information they bring gives us a detailed account of where they've come from, their birthplace. It's as if the Universe is communicating with us through light. It's telling us about itself.

Now, the microwave background is our current example. You can think of it as an ancient panoramic document, onto which is written the story of the Universe's youth. Those patterns of patches are like a wall of hieroglyphs, and our task is to read them in order to learn their story. So what language is this document written in? I mean, certainly, it's not any human language, but it is, in fact, written by Nature, in Nature's own language. Now, you may not have thought that Nature has a language, but it does—it's physics. So to understand what Nature is telling us, one needs to speak physics. It's quite intriguing that our brains are, in fact, able to learn this alien language, though for most of us, it's at least as difficult as learning ancient Greek or Latin. But fortunately, there are those who are experts, and they've been poring over all those hieroglyphs, trying to figure out what they mean. That's our story for this lecture: how these experts approach deciphering this primeval document, written by Nature, back near the dawn of time.

Now, it comes as a huge relief to discover that much of the physics of the young Universe is surprisingly simple. In a sense, the young Universe is sort of speaking in a youthful, simple way, and there are four reasons for this simplicity:

- The young Universe is almost perfectly homogeneous. It's just a smooth gas.
- The components of this gas are themselves relatively simple and well understood: light, electrons, nuclei, dark matter. And furthermore, the light, electrons, and nuclei are all tied tightly together into a single, coherent photon-baryon foggy gas, so for much of the time there are really just two components—photon-baryon gas and the dark matter.
- In addition to being tied together in a single fog, the light, electrons, and nuclei are also in thermal equilibrium, and to a physicist, this is the simplest kind of situation one could have. It's the kind that undergraduate textbooks routinely deal with.
- The deviation from exact homogeneity—in other words, the lumpiness—is very slight, and the physics of how slight lumpiness grows during expansion is relatively straightforward. Physicists say the situation is in the linear regime, which means it yields to a particularly simple mathematical treatment.

Now, having said how simple the young Universe is, it still took almost 30 years—from the 1970s to the 1990s—for theoretical cosmologists to construct a detailed treatment of the hot, expanding gas laced with slight fluctuations. And some of the key contributors were Andrei Sakharov (the famous dissident), Rashid Sunyaev, and Yakov Zeldovich in the former Soviet Union; and Jim Peebles, Joe Silk, George Efstathiou, and Dick Bond in the U.S., UK, and Canada; though of course many others were also involved.

Now, turning to the second step in the overall project, knowing the theory isn't enough. One needs to actually calculate the primordial sound and its appearance on the microwave background. These calculations demand highly sophisticated computer programs run on powerful workstations, and the development of these programs has been a huge task involving many people, taking over 15 years to complete. Again, some of the key contributors were Ed Bertschinger, Naoshi Sugiyama, Uros Seljak, Matias Zaldarriaga, and Wayne Hu.

Now, perhaps the most famous computer program goes by the name CMBFAST, and as its name suggests, it uses some very clever methods to reduce the calculation time from days to seconds. Running these programs has become so important to cosmology that NASA has sponsored a public Web interface as part of its LAMBDA project, so that even *you* could simulate your very own young Universe. I'm not kidding. It's really very easy. Although I don't have time to go over the details, you don't need to worry about them because the default values all work just fine. So let me take you through how one calculates a young Universe.

There are actually three stages to the program. First, you must specify the basic parameters of the Universe you want to calculate. So on the Web interface, this includes things like the Hubble constant and the relative amounts of atomic matter, dark matter, dark energy, and so on. There's also an important parameter we'll discuss in the next lecture that specifies the initial seed lumpiness emerging from the Big Bang itself—the set of organ pipes, if you like. With these now specified, stage two is running the program. In the 10 or so seconds the program runs, an incredibly complex set of calculations is being done. The lumpiness within each component is calculated as a sound spectrum, as it responds to the pressure and gravity of all the other components.

This sound spectrum is then tracked forward in time from the Big Bang. The calculations track the true variations in pressure and density, what in the last lecture I called the "pure sound spectrum," $P(k)$. You may recall this has a relatively clean set of harmonics, shown here in red, at the time near 400,000 years. But remember, we don't see these waves directly—we only see the patches on the microwave background. So the third stage in the calculation is to follow the waves through fog clearing and calculate how they appear as patches on the microwave background. Stated a little more technically, we need to go from $P(k)$ to $C(\lambda)$, which is the angular sound spectrum of the patches seen on the microwave background, shown here as the blue line.

This is a crucial step and includes some very interesting processes that I think it's worth taking some time to explain. Let me set the stage with a simple metaphor. As we look out at the microwave background and record its sound, it's not unlike sitting way at the back of a rather bad concert hall with poor acoustics. Now, the true sound is deep in the gas. That's the orchestra playing the symphony down on the stage. But from where we're sitting, the concert hall has carpet and drapes that deaden the high notes, a ceiling that muddies the low notes, and our neighbors constantly rustle candy wrappers. So the symphony we hear has all these other effects added in. In the same way, you can think of the sound we measure from the microwave background, $C(\lambda)$, as a distorted version of the true sound, $P(k)$.

Now, pushing our metaphor a little further, if you already know what the symphony should sound like, then the distortions actually allow you to figure out quite a bit about the nature of the concert hall. So we anticipate learning even more about the Universe from the presence of these distorting effects. Now, in reality, CMBFAST must deal with about 8 or 9 of these distorting effects, but I'm just going to mention 2 or 3 of the more important ones. So let's begin with one that is relatively straightforward. Remember how the sound waves are being formed: We have slightly over-dense pockets of dark matter surrounded by under-dense regions. So gravitationally, it's as if there's a valley surrounded by hills, and the gas feels a pull into the valley. The gas falls into and then bounces out of these dark matter valleys, creating higher pressure and temperature, which makes the gas glow brighter.

Well, if you think about it, while the gas is moving, either falling in or bouncing out, the light it emits is Doppler-shifted. Gas moving towards us looks a little bluer and brighter, while gas moving away looks a little redder and dimmer. So the motion of the gas makes its own patches on the microwave background. If you think very carefully about this, you'll see that the patches of maximum motion have intermediate size between patches of maximum compression and patches of maximum rarefaction. And so the effect of adding patchiness on these intermediate scales fills in the gaps between the harmonic peaks, and that's why $C(\lambda)$ has much less prominent harmonics than $P(k)$.

The second effect was one of the first to be discussed back in the mid-1960s by Rainer Sachs and Arty Wolfe, and it's called, perhaps not surprisingly, the "Sachs-Wolfe effect." It's really very simple. Imagine a photon coming from the middle of a region of excess dark matter. As the photon leaves the region, it has to sort of climb uphill to get out of the valley, where "uphill" means against the pull of gravity. But climbing uphill drains some of the photon's energy, so it gets redder. This is called a "gravitational red shift," and it means that over-dense regions will appear slightly darker on the microwave background because their light has been redshifted and so dimmed a little bit.

Something that's important about the Sachs-Wolfe effect is that it happens for regions of all sizes, including huge regions like the ones shown here. Now, remember that the acoustic oscillations only occur for regions up to, but no bigger than, the sound horizon. Basically, there hasn't been enough time for gas to fall into and bounce out of yet-bigger regions. Now, on this plot of the sound spectrum with an exponential x-axis—which makes it look a little different than the previous ones I've shown you—the boundary is very obvious, with all the acoustic and Doppler peaks to the right of the sound horizon, which is about $2.0°$ across on the sky. In fact, that whole region to the left is called the "Sachs-Wolfe plateau." In our metaphor, it's the deep hum of the air conditioners and is unrelated to the sound of the symphony.

Going back to our boundary between the small acoustic patches and the large Sachs-Wolfe patches, you may not have realized that most of what your eye sees in those lovely microwave background images is, in fact, the Sachs-Wolfe effect, not acoustic waves. Here's a diagram showing a whole range of patch sizes—first across a quarter of the entire sky, and then across a small $10° \times 5°$ subregion, with the range of patch sizes between $2.0°$ and $\frac{1}{10}°$ shown as little circles. Now, here's where those patch sizes end up on our sound spectrum. You can see that even the largest acoustic waves, which are not much bigger than the full Moon on the sky, are tiny when you place them on these grand all-sky maps. You can also see why it's so important to have high angular resolution. The WMAP image here is not really adequate to probe the higher harmonics, which is why the next satellite, called "Planck," is designed to have much higher sensitivity and resolution.

So let's now consider these smaller waves. They appear over on the far right of the sound spectrum, and you can see how they seem to have been suppressed. In our concert hall, this is the equivalent of carpets and drapes deadening the high pitches. For the microwave background, there are actually two effects responsible for this apparent suppression of small waves. Let's have a look at those now.

The first effect arises because the wall of fog on which we see the sound waves is itself somewhat fuzzy. It has depth. Remember, the wall arises from us looking out into space, back in time, through the period of fog clearing. But since it took thousands of years for the fog to clear, then the boundary between fogginess and transparency is itself thousands of light-years deep. So if a sound wave is smaller than this thickness, then both the peak and trough are within the wall's thickness, and they get blurred together, and so the wave appears washed out. We don't see it. And that suppresses the sound spectrum.

Now, the second effect arises because, despite what I've told you before, the Universe isn't completely transparent. There's a misty curtain lying just in front of the microwave background, which blurs the patchiness and suppresses the sound spectrum. Here's what's going on. As we learned in Lecture Thirteen, 400,000 years after the Big Bang, the Universe turned from foggy to transparent as it turned from ionized gas to atomic gas, and at that point the Universe began a long period of ever-colder darkness that's come to be called the "dark age." Now, we'll visit the dark age in Lecture Eighteen, but for now, just know that after about 400 million years, the first stars were born, and their powerful light began to reionize the neutral gas, reversing, in a sense, what had happened at 400,000 years. Now, this ionized gas introduces a thin mist, and so we actually see the microwave background through this mist. And as you might imagine, this blurs its appearance and tends to make the smallest patches less visible, so the high-frequency sound waves appear to be suppressed.

Now, you might wonder why ionized gas at 400 million years is only a thin mist while ionized gas at 400,000 years is a dense fog, and the answer is that by 400 million years, cosmic expansion has made the gas so much thinner that it's only a misty curtain, rather than an impenetrable fog. Okay, look, that's enough of these complicating processes. Really, don't worry if you didn't quite follow them all. Just recognize that they are sufficiently well understood to be included in the final stage of the CMBFAST calculations, which brings the simulated young Universe right to the point that it can be compared with data. In other words, it calculates the sound spectrum $C(\lambda)$. Here, for example, is the calculated sound spectrum for the model I showed you earlier. Now, if you run your own models, this plot, along with other things, is what's returned to you when the program has finished its calculations.

Now we come to the third and final part of our primary project, which is to use the sound spectrum to measure the properties of the Universe. We try to find a set of input parameter values for CMBFAST that result in a calculated sound spectrum that matches the observed sound spectrum. Now, this time, I'm not going to focus on the actual derived parameters, because we're going to use Lecture Twenty-Six to discuss a much broader program to match cosmological models to 4 or 5 quite different cosmological data sets, only 1 of which is the microwave background. But here, let's just concentrate on seeing to what extent CMBFAST can actually match the observed sound spectrum.

Now, someone who's worked quite hard on this kind of model fitting is Max Tegmark, and I'm going to be using some of his movies in a moment. So the simplest way to illustrate all this is to vary just one parameter at once and watch the change in sound spectrum. It's a bit like playing a set of violins, all identical except for the type of wood they're made from. Here's one of Max's movies, showing how the calculated sound spectrum changes as we increase the total cosmic density. Clearly, as the total density increases, the entire sound spectrum shifts to lower frequencies. Denser Universes have deeper voices. Look at how the data really single out a specific density as the true one. The real Universe has a well-defined pitch, and this pitch tells us within about 1% accuracy what the total density is.

Now, just for fun, let's listen to the primordial sound of 3 Universes spanning a range of 40% in total density. The change in pitch is really obvious, and the middle one with the black line matches the observed data best. Now, that was the total density. What about the fractional density of different components? Here's how the Universe's sound spectrum changes as you change the relative amounts of atomic and dark matter from no atomic (that's all dark) to all atomic (that's no dark). You can see how the first peak gets much stronger, while the location and strength of the higher harmonics change more subtly. Again, the data clearly pick out an atomic content near 4% and dark matter content near 23%.

As before, it's fun to see what these different Universes sound like. Here are 3 models with atomic matter content set at 8%, 4%, and 2%, and I've plotted the fundamental peak at the same height so that you can see more clearly the difference in the higher harmonics. That's actually a nice example of how the sound spectrum easily picks out the difference, but it's actually quite difficult for our ears to hear the difference.

Now, my last example is to vary the time at which the first stars form. Remember, they reionize the gas and create a foreground mist through which we see the microwave background. With no foreground mist, the model lies above the data, but as the mist gets thicker and thicker, the microwave background appears more and more blurred and its patchiness gets washed out, so the measured sound spectrum gets weaker and weaker. The data suggest the mist isn't very thick, about 92% transparent, which corresponds to a time for the birth of the first stars around 400 million years, with an uncertainty of about 30%, which if you think about it, is a really quite remarkable measurement.

Well, let's bring this lecture to a close by returning to our graph, showing the best microwave background data and the best model fit as of about 2007. The model has a total of about 8 parameters, many of which are determined to within a few percent accuracy. As I mentioned before, we'll postpone to Lecture Twenty-Six a fuller discussion of all of these parameters when we bring in all the other data sets. But for now, I want you to step back and let the full message of this diagram hit you. It's far too easy to get lost following the details and miss the stunning nature of this plot.

First, the observations: Just look at the tiny size of those error bars, at least over the first two peaks. And when the results from the next major satellite are added, the European Planck Surveyor, they will be similarly small across the entire plot and beyond. This is a fabulously exact measurement of the acoustic structures present in the young Universe, and there is no mistaking the wonderfully rich spectrum, with its harmonics, low-frequency plateau, and high-frequency damping. But even more impressive, I think, is our ability to model it so accurately.

This is not some broad, featureless spectrum that might be explained by all kinds of general mechanisms. No, it has multiple peaks of specific height and location, and our model gets them all correct, passing right through the data points. I think this plot, more than any other I know, should convince you that we're on the right track, that what I've been telling you about the first million years, with its foggy brilliance, its endless sea of dark matter, its deep, surging sound waves, is not just a modern myth. It all actually happened, long, long ago, before the Sun and Earth had formed, and well before you and I were here to wonder how it all began.

Lecture Seventeen
Primordial Roughness—Seeding Structure

Scope:

How does the Universe evolve from being extremely smooth to being filled with stars and galaxies? The next six lectures follow this 14-billion-year transition. We begin by focusing on "roughness"— slight variations in density from place to place that were injected by inflation and grow over time. To understand this growth, we introduce the concept of the "roughness spectrum," which specifies the relative mix of long, medium, and short waves in density. Whether a wave is smaller or larger than the horizon size affects how its roughness grows. In particular, waves smaller than the horizon cannot grow in roughness during the radiation era, and this introduces a peak in the roughness spectrum—a scale of maximum texture. Observations of the real Universe wonderfully confirm the theoretical roughness spectrum, complete with its peak—a relic of the radiation era written in the pattern of galaxies.

Outline

I. This fourth theme traces the growth of structures: from the smooth young Universe, to the first stars and infant galaxies, to maturing galaxies, to the galaxy web.

    A. Throughout, there are two key players: atomic matter and dark matter. Each has its role to play, and each by itself couldn't create today's rich Universe.

    B. Overall, somehow, the Universe changed from being very smooth to being very lumpy. How did it do that?

    C. This lecture looks at the beginning of that process and asks three questions.

        1. How did roughness start?

        2. How does roughness grow?

        3. Why does the degree of roughness depend on scale?

II. How did roughness start?

    A. Perhaps the Universe was born perfectly smooth, and roughness grew from nothing? There is no known mechanism that can do this.

B. The theory of inflation includes a quantum process that naturally sows the seeds of roughness. This is how we think roughness started.

III. The nature of roughness.

A. Roughness is defined to be the ratio of slight variations in density to the average density: $\Delta\rho/\rho$. A 6-foot pool with 6-inch waves has 8% roughness.

B. During expansion, roughness grows even though density drops.

C. We're going to need two important concepts: the horizon size and the roughness scale.

D. The horizon size is the distance light (or gravity) can cross since the Big Bang (e.g., at 100 years it's roughly 100 light-years).

   1. Everything within a horizon distance is causally connected.

   2. Everything outside a horizon distance is utterly independent.

   3. The horizon size gets bigger at light speed.

E. Roughness occurs on many scales.

   1. For example, a particular density pattern may have long undulations with short ones superposed. There are two "wavelengths" present.

   2. The wavelength defines the scale of the roughness component.

   3. Such waves don't oscillate in time; they're fixed in space with wavelike form.

F. Depending on whether a wave is longer or shorter than the horizon, it undergoes super-horizon growth or sub-horizon growth.

G. Because the horizon gets bigger at light speed, all waves start with super-horizon growth and then change to sub-horizon growth.

IV. Super-horizon roughness growth.

A. Denser and less-dense regions separated by more than the horizon behave like independent Universes.

B. Denser regions expand more slowly than less-dense regions, so the density contrast (roughness) increases.

C. This is extremely potent: From a nanosecond to a thousand years, roughness grew by 10^{20}.

V. Sub-horizon roughness growth.

    A. Neighboring regions within the horizon feel each other's presence.

    B. Denser regions pull material from less-dense regions, so the density contrast increases.

    C. Isaac Newton recognized that a static Universe was unstable to this kind of clumping, which would proceed at an accelerating pace.

    D. However, cosmic expansion dilutes the matter, and this weakens the effect.

        1. In the radiation era, expansion is too rapid, and clumping cannot occur.

        2. In the matter era, expansion is less rapid, and roughness can grow.

    E. The fact that sub-horizon growth is suppressed in the radiation era will turn out to be very important, so hold that thought.

VI. Our final prep work is to describe roughness more fully. Imagine a graph of roughness across the Universe: a complex line going up and down. We can make that line by adding together many waves of different wavelength, each with a particular strength.

    A. The graph of wave strength, P, against wave frequency, k, we'll call the "roughness spectrum," $P(k)$. (Cosmologists call it the "power spectrum.")

    B. The wave frequency, k, is low on the left (long waves) and high on the right (short waves).

    C. We met $P(k)$ in Lecture Fifteen on Big Bang acoustics. There it described pressure variations in the glowing gas. Here it describes density variations in the dark matter. It's a very similar concept.

    D. In summary, *any* roughness graph can be recast using a roughness spectrum: the set of waves that, when added together, make that graph.

    E. Joseph Fourier invented this kind of dual description in the 19th century.

VII. We're now ready to track the growth of cosmic roughness from the Big Bang to today.

   A. The initial roughness injected by inflation has a spectrum $P(k) \propto k$, so it's smooth on large scales and rough on small scales.

   B. Following how this spectrum changes over time is quite tricky. A first sketch has time along the bottom and size upward.

   1. Consider 5 density variations, from short to long wavelength. Cosmic expansion stretches them all, from bottom left to top right.

   2. Because the horizon size gets bigger at light speed, it envelopes each wave in sequence, from short to long.

   3. The 2 shortest waves enter the horizon during the radiation era.

   C. Now let's consider the growth of roughness for these 5 waves. A second sketch replaces the *y*-axis with roughness, $P(k)$.

   1. The initial roughness (on the left) increases from weak, long waves to strong, short waves.

   2. The roughness of the longer waves grows continuously.

   3. The shorter waves enter the horizon during the radiation era, and so their roughness freezes. When the radiation era ends, their roughness starts growing again, but they end up less rough than the longer waves.

   4. Today's roughness (on the right) *increases* from long waves to medium waves, but then *decreases* to short waves.

   D. A full calculation using CMBFAST shows how $P(k)$ evolves: The initial spectrum moves upward but acquires a peak near the horizon size. At the end of the radiation era, the peak stays put, and the whole spectrum just shifts vertically upward as roughness grows equally on all scales.

   E. This peaked roughness spectrum generates an interesting density pattern.

   1. It is quite chunky on a scale matching the peak of $P(k)$.

   2. The density peaks correspond to huge superclusters of galaxies.

   3. The density troughs correspond to huge voids.

   4. The small-scale roughness corresponds to galaxies.

VIII. Let's end this (difficult) lecture with the real Universe's roughness spectrum.

    A. To measure it, we need to combine two kinds of data sets.

        1. Galaxy red shift surveys map roughness out to scales ~1 Bly.

        2. The CMB patches give the roughness on larger scales.

    B. The overall shape of $P(k)$ is just what we expect. From low to high k:

        1. It increases proportional to k.

        2. It passes through a broad peak.

        3. It falls, more steeply, to smaller k.

    C. The peak sits at about 200 million light-years. It's the scale of maximum texture.

        1. It arises because sub-horizon roughness couldn't grow during the radiation era.

        2. Its scale corresponds exactly to the horizon size at the radiation/matter transition: 60,000 light-years stretched by 3700 become about 200 Mly.

        3. It is remarkable that this transition, way back in the Universe's infancy, should leave its imprint on the patterns of galaxies across the sky.

Suggested Reading:

Hawley and Holcomb, *Foundations of Modern Cosmology*, chap. 15.

Silk, *The Infinite Cosmos*, chap. 13.

———, *The Big Bang*, chap. 9.

Questions to Consider:

1. A crucial part of cosmic history is the amplification of very slight roughness introduced during inflation. How does this roughness grow? Be careful to distinguish between super-horizon growth and sub-horizon growth, because these are radically different mechanisms.

2. Although the roughness spectrum injected by inflation takes the form $P(k) \propto k$ (weak, large wave roughness and strong, small wave roughness), the roughness spectrum today has a peak for waves around 200 million light-years and then gets *less rough* for smaller waves. Why is this? Why have the smaller waves of roughness not been amplified as much as the larger ones?

3. Why does the peak in the roughness spectrum correspond to the (cosmically stretched) horizon size at the end of the radiation era? How does this peak show up in the distribution of galaxies—what features do you expect to see in their patterns that correspond to this peak?

Lecture Seventeen—Transcript
Primordial Roughness—Seeding Structure

This is the first of six lectures that will make up our fourth theme, where we'll trace the growth of structure, following a sequence from the Big Bang, through sound, to stars, to galaxies, to their weblike distribution. Here's a brief preview of the path we'll take. There are two key players in the story, atomic matter and dark matter, and in many ways, they act like a team. Each has its vital role to play, and neither by itself could create the beautiful Universe we find around us.

I could almost assign them human roles. Dark matter might be the parent, and atomic matter the child. Now, without the parent's guidance, the child wouldn't flourish, but what the child ultimately does, the things it makes and the games it plays, are its own creation. We, of course, living in our atomic world, are the child. But we should never forget that we wouldn't be here if it weren't for our parent, dark matter, which gently provides a context, a framework, for our atomic world to form and grow.

So the first lecture in this theme, this lecture, looks at how dark matter makes that framework into which atomic matter will ultimately gather. We start with seed roughness, created in the Big Bang, and look at how it grows rougher over time. The second lecture walks through the first billion years. Regions of slightly higher density finally halt their expansion, turn around, and collapse to form mini-halos of future galaxies. Meanwhile, atomic matter falls into these dark matter mini-halos and continues collapsing in a runaway process that ends in the birth of the first stars.

Then the third, fourth, and fifth lectures concentrate on this atomic matter, as neighboring infant galaxies collide and merge to form ever-larger systems, ultimately becoming the present-day galaxies that we see all around us. And then in the last lecture, we'll return to look at the large-scale structure of dark matter, the weblike framework revealed by the distribution of thousands of galaxies. This is a dynamic, evolving framework, within which atomic matter is trapped and lives out its creative life—so creative, in fact, that of course it makes stars and planets, rocks and trees, and you and me.

This fourth theme has many wonderful things to explore, so let's get started. First, I want to frame the big picture. Remember, the

Universe is born very smooth indeed—an endless glowing gas. The microwave background, for example, shows us this gas is uniform to 1 part in 100,000. Now, meanwhile, 14 billion years later, we live in an amazingly lumpy Universe. There are vast voids, with almost no galaxies in them. There are some patches with more galaxies, and some regions absolutely crammed with galaxies. It's like a giant 3-dimensional patchwork, mottled full of denser and less-dense regions, going on and on, filling the Universe.

So, stated simply, the question for this lecture is, "How did we get from so smooth to so lumpy?" It's a bit like one of those *Just So Stories* by Rudyard Kipling, "How the Leopard Got His Spots" or "How the Camel Got His Hump," and in this case, it's how the Universe got its lumps. Now, there are going to be three parts to the story. The first asks, "How did roughness start?" Can it arise naturally out of perfect smoothness, or must the Universe be born with roughness? The next step is to understand how roughness grows. Why do variations in density get larger as time goes by? And lastly, we need to understand why the degree of roughness depends on size. From one kilometer to the next, the ocean is very smooth, but from one meter to the next, it's very rough. The Universe is similar. Why is that?

Okay, let's start with the first of these questions—how did roughness start? Now, one possibility is that the Universe was born perfectly smooth, and roughness arose out of nothing—naturally, perhaps, by some kind of instability. Well, in the 1960s and '70s, a lot of thought went into this, and every possible mechanism that was considered wouldn't work. If the Universe was born perfectly smooth, it would stay perfectly smooth. And then in the 1980s, as people began to work on the new theory of inflation, it came as a wonderful surprise to find that roughness could be generated quite naturally from tiny quantum mechanical fluctuations that get amplified during the inflationary launch of the expansion. We'll look at exactly how this mechanism works in Lecture Thirty-Two, but the bottom line is that the Big Bang itself makes the initial roughness.

So having postponed answering our first question, let's consider the second one—how does the initial seed roughness grow? Let's start by defining roughness, using this graph. Imagine moving through the early Universe, along a line. The density of matter varies a little bit from place to place. It gets denser and then less dense above and below the mean, in a haphazard way like this. The roughness is defined to be the ratio of the variation to the mean. In math terms, the roughness is $\Delta\rho/\rho$, where ρ is the average density, and $\Delta\rho$ is the small variation. Think of a swimming pool 6 feet deep, with 6-inch waves. Its roughness is 1/12, or about 8%.

In this lecture, what we're interested in is how this roughness grows. The average density, of course, drops with expansion, but somehow superimposed on that, the contrast between denser and less-dense regions gets more pronounced. Now, as the Universe expands, there are, in fact, two rather different ways in which the roughness grows. One is called "super-horizon growth," and the other is called "sub-horizon growth." So I need to explain a little bit about what these rather impressive-sounding terms mean.

We've come across the concept of the horizon before, but let's just review it. Out on the ocean, you can see to your horizon. You have knowledge of, you might even say you're affected by, everything within your horizon, but not beyond. And it's like that in the early Universe. Every location has a sphere of influence, a cosmic horizon. It's a region across which light or gravity can cross and influence things. Its size is roughly the distance that light can travel since the Big Bang. So 10 years after the Big Bang, your horizon is about 10 light-years across; 100 years after the Big Bang, it's about 100 light-years across.

Now, like the ocean, every point has its own horizon, its own sphere of influence, but unlike the ocean, the cosmic horizons get bigger at light speed. Now, the reason that the horizon is important for the growth of roughness is that roughness occurs across different length scales. Again, a bit like the ocean, the surface has small waves and long waves, and whether a wave is larger than the horizon or smaller than the horizon affects how it can grow.

Here, for example, is the variation in density above and below the average, along a line spanning about 1000 light-years. It clearly contains two rather different roughness scales, one short and one long, each shown below. The top curve is basically the sum of the lower two. So when we speak of "roughness length scale," we're really referring to the wavelength of a particular component that's present in the total. See, in this simple example, we only have two components, and so you can see that their wavelengths are about 120 light-years and 8 light-years. Notice I used the word "wave" as in "wavelength" to refer to the up and down quality of the variation, even though it isn't actually oscillating in time, only in space.

Now, let's compare these two different wavelengths to the horizon. Imagine it's just 30 years after the Big Bang, so the horizon size is about 30 light-years, the size of these little circles. So clearly at this time, the large wavelength roughness is larger than the horizon size, while the small wavelength roughness is smaller than the horizon size. This is very important because waves smaller than the horizon and waves larger than the horizon experience two different kinds of growth. Hence we speak of "sub-horizon growth" for the waves smaller than the horizon, and "super-horizon growth" for waves larger than the horizon.

Okay, one last point. The horizon is, of course, getting bigger at light speed, and so before too long, the horizon size catches up and envelops the larger waves. And when it does so, we say these waves enter the horizon. When they do this, they change, of course, from being super-horizon to sub-horizon, and so the way in which their roughness grows also changes. Let's now have a look at super-horizon roughness growth. It's actually very simple.

Here are three regions. A is slightly higher density, B is average, and C is slightly lower density. Notice that they're all way bigger than the horizon, which has the size of this small circle. So, and this is the important part, these three regions are completely unaware of each other, and they expand just as if they were three different Universes with three slightly different densities. Now, here's a sketch of how their size grows with time, their expansion curves. Just as you'd expect, the denser region, A, is lagging behind because it's held back by the pull of the extra density. As time passes, the difference in region sizes gets bigger, so the difference in region volumes gets bigger, so the difference in region densities gets bigger, so the roughness increases.

Basically, the density contrast grows because the denser regions aren't expanding as fast as the less-dense regions. Now, this is an exceedingly potent amplification process. For example, during the radiation era—that's before 60,000 years—density variations grow in proportion to time, so from a nanosecond to 1000 years, the roughness grows by a factor of nearly 10^{20}. Now you can see how miniscule seed fluctuations emerging from inflation can end up becoming significant at later times.

Okay, now, let's look at sub-horizon growth. Here's the same situation as before, but with our three regions, A, B, and C, now well within the horizon size. In this situation, each region feels the gravitational pull of the material within a horizon's distance. So there are local gravitational forces, where denser regions actually pull material in from nearby, less-dense regions, and so they get denser. And meanwhile, because the less-dense regions are losing material to the denser regions, they get even less dense. So once again, for completely different reasons this time, the roughness increases.

Now, this particular process actually has a long history. Isaac Newton thought deeply about how his theory of gravity might apply within an infinite Universe, and in a famous dialogue with the theologian Richard Bentley, Newton realized that an infinite, static Universe would gradually clump up. Places where the density was even a tiny amount more would pull material in from nearby, and so its density would grow, ultimately, in a runaway, accelerating collapse. In fact, he even wondered whether stars might have formed that way.

Now, in our situation, there is one very important difference. Our Universe isn't static—it's expanding. Now, this expansion continually dilutes the matter, weakening its gravitational pull, and this seriously affects the ability of roughness to grow by this mechanism. In fact, it's only during the matter era, after 60,000 years, that the expansion is sufficiently gentle to allow some growth in roughness. It turns the runaway collapse for a static Universe into a slower growth proportional to the scale factor. So if the Universe doubles in size, then the roughness doubles in size.

Now, remember from Lecture Twelve that the expansion during the radiation era, before 60,000 years, is more rapid and, in a sense, more potent, and this basically prevents the growth of sub-horizon roughness. We say the roughness gets frozen at whatever level it was when it entered the horizon. So sub-horizon growth of dark matter roughness only occurs during the matter era, not during the radiation era. And this will turn out to be extremely important, so hold that thought because we have one more piece of prep work to do before we put all this together and understand how the Universe got its lumps.

This prep work is to understand how cosmologists describe roughness. Going back to our example from before, it turns out that an extremely useful way to think of any complicated pattern of roughness is as the sum of simpler patterns of different wavelength. In this case, there are really just two of them—short waves and long waves—which combine to make the top curve. But now look at this complex set of density variations, again spanning 1000 light-years. Well, it turns out that this upper curve can be made using the lower curve, which I'm going to call its "roughness spectrum," although the technically correct term is its "power spectrum." The idea is very similar to a sound spectrum from Theme III. The red line shows the strength of waves of different wavelength that are present in the upper curve. Notice for the x-axis, we have long waves (small frequencies) on the left and short waves (high frequencies) on the right.

Remember, these waves don't oscillate in time. They're fixed in space, with a wavelike form. So the frequency is really a spatial frequency, meaning the number of waves per 1000 light-years, and it's labeled k, while the strength of each wave is labeled P for power. So this curve is called $P(k)$. Now, you should remember this term from the last couple of lectures on Big Bang acoustics, where I referred to $P(k)$ as the pure, or true, sound spectrum. Now, in that case, variations in the glowing gas were of pressure, whereas in this current context, the variations are in density, and the density of dark matter as well. But either way, you can think of $P(k)$ as showing the strength of waves of different wavelength.

So the remarkable fact is that you can take a whole set of waves, all with different wavelengths, but with strengths given by this red curve, and you can add them all together, and you can get the top curve. It's really quite amazing. In fact, you can make any graph this way. It doesn't matter what shape it has. You can always find a set of waves that, when added together, make that graph. The person who first figured out how to do all this was in the 19th century, a French mathematician called Joseph Fourier, and so these two complementary ways of representing a graph are called "Fourier pairs," and the methods used to calculate one from the other are "Fourier transforms." They're really amazing things and are widely used by physicists and engineers.

Okay, we're now done with preparation, so we're ready to put all of this together into a single story, which tracks the growth of cosmic roughness from the Big Bang to today. Our starting point must be the roughness injected by inflation. Now, in Lecture Thirty-Two, we'll find out that inflation generates a roughness spectrum—what cosmologists call the "initial power spectrum"—that is particularly simple. It's just $P(k)$ proportional to k, so the roughness strength is proportional to the wave frequency.

Here's a plot of this kind of roughness spectrum, along with its corresponding density pattern. On very large scales—that's small k on the left—the waves are very weak, so it's very smooth across gigantic regions, a bit like our ocean. Viewed from an airplane, it looks smooth. This is just what we expect. After all, the Universe is homogeneous—it's smooth—on the largest scales. Conversely, on small scales, that's large K, the waves are strong—so the newborn Universe was much rougher on finer scales. Okay, armed with this initial roughness spectrum, let's now follow its evolution over all of cosmic history. In other words, let's follow how those initial tiny fluctuations grow to become today's weblike pattern of galaxies.

Now, the next two diagrams are very important, so try to follow along. Here's a sketch with time along the bottom and size upwards. Both axes are exponential, but we're not going to bother with details of actual numbers, but two times are shown—today, and the transition between radiation and matter eras near 60,000 years, which I've labeled "equality." On the left are 5 density variations, with wavelengths of increasing size, 1 through 5. Cosmic expansion makes all of those wavelengths expand, following the yellow lines to the right. They all get stretched in size by the same expansion factor, and arrive at their present-day sizes at the top right. Now, the red line shows the increasing horizon size. Remember, it gets bigger at light speed, and where it crosses each yellow line, the horizon envelops that wave, and the wave enters the horizon at the five times shown by these little circles. Notice that the 2 shorter waves enter the horizon during the radiation era. The 2 longer waves enter the horizon in the matter era, and the middle wavelength enters the horizon right at the time of equality.

Okay, now, let's consider the growth of roughness for these 5 waves. In this graph, the y-axis is now roughness, measured by $P(k)$, the strength of those 5 waves; and time is still along the bottom axis. At early times on the left, the strength of the 5 waves is given by the initial power spectrum, $P(k)$ proportional to k, the frequency. So the roughness increases from weak long waves—that's number 5—to strong short waves—that's number 1. Starting with the longer waves, 5, 4, and 3, they all grow continuously. They remain outside the horizon throughout the radiation era, and so their growth is never frozen. And when they enter the horizon in the matter era, they continue to experience sub-horizon gravitational growth. But the shorter waves, 2 and 1, enter the horizon during the radiation era, and so their roughness, $P(k)$, gets frozen at that time. They stop getting rougher.

Now, as soon as the radiation era ends, their roughness starts growing again, but they can never make up for the lost time, and they end up with less roughness than the longer waves. So although the initial roughness—on the left—steadily increases, 5, 4, 3, 2, 1, the final roughness first increases, 5, 4, 3, reaches a peak at 3, and then decreases again, 3, 2, 1. Let's now see how the roughness spectrum evolves over time. Effectively, we'll move up these green lines, and we're going to ignore the simple cosmic expansion of the wave sizes, so we can concentrate on the roughness spectrum alone. We'll also go beyond the cartoon treatment I've been showing you, and we'll use instead a full calculation, using the CMBFAST computer program from the last lecture.

Here's a graph with wave frequency k along the bottom and wave strength $P(k)$ upwards, with both axes exponential. You can see how the roughness spectrum increases over time, starting at 100 years, then 1000 years, then 10,000 years, and so on. So we're jumping by factors of 10 up to today, at 10 billion years. So at just 100 years, you see mainly the primordial spectrum, $P(k)$ proportional to k. But as time passes, you can see how the strength of the small waves gets held back when they enter the horizon during the radiation era. But when that era ends, all waves then grow in strength together, so notice that a peak forms, close to the size of the horizon, with the primordial spectrum to the left, and the suppressed smaller waves to the right. The peak gradually shifts to larger-length scales during the radiation era but settles down to the horizon size at the end of the radiation era, and from then on, the roughness spectrum just shifts vertically upwards, as roughness grows equally on all scales.

So today, after 10 billion years, the Universe's roughness spectrum should have a peak whose position is a relic of the transition from radiation to matter dominance around 60,000 years after the Big Bang. Let's see how this roughness spectrum converts to a pattern of density variations along a line. Here's the spectrum, with peak and suppressed high frequencies, and in the density plot, you can see how all the very high frequency waves are now gone, and we have quite a chunky graph, with the strongest variations having a scale corresponding to the peak in the spectrum. Now, those density peaks will ultimately turn into huge superclusters of galaxies, while the troughs will turn into enormous voids, and the smaller ups and downs will make individual galaxies.

Well, at this point, we really must turn to the real Universe to see if all this effort was worth it. In fact, before I do that, I'd just like to acknowledge that this has been a very tough lecture, which probably needs listening to more than once. I deliberately chose to aim high and take you through a topic that is rarely, if ever, presented to nonspecialists. So if you've followed even a portion of it the first time through, you should feel pleased with yourself. We're on the home stretch, about to look at the real Universe's roughness spectrum, and hopefully it should look familiar.

Now, to span the full range of scales, we need two different kinds of data sets. The first are the huge surveys of hundreds of thousands of galaxy positions that reveal the roughness, spanning a range of scales from a few million light-years to a few hundred million light-years. Your eye can already see that the strongest contrast is on fairly large scales, perhaps a few hundred million light-years. And sure enough, when you take the Fourier transform, you find a roughness spectrum, $P(k)$, rising for larger scales, shown here as the blue data points. Now, to get data on even larger scales, one must use the patches on the microwave background.

These, of course, apply to the young Universe, so they need to be adjusted to bring them forward to today's time, but when you do that, you get the red data points, which nicely overlap the galaxy data set, and the overall shape is clear. It has just the form we now understand. From left to right, on the larger scales, the roughness spectrum rises proportional to k. It then passes through a broad peak, corresponding to a length scale of about 200 million light-years. It then falls somewhat steeper on smaller scales. The black line comes from running one of the best CMBFAST computer models. The fit is amazingly good, right down to the little wiggles on the right.

Now, we'll come back to those little wiggles in Lecture Twenty-Two, but let's end this lecture by reflecting on what we have in front

©2008 The Teaching Company.

of us. We've measured the texture of the Universe, from 30 million to 6 billion light-years. In a sense, this is the inherent large-scale texture of reality, not the particles or atoms of the microworld, but the scaffolding and stage on which the entire play unfolds. Remarkably, there is a scale of maximum texture, about 200 million light-years, shown by the peak in the roughness spectrum. It arises because the growth of small-scale roughness was suppressed or held back on sub-horizon scales during the radiation era. If we take the horizon size at that time, about 60,000 light-years, and we apply the appropriate cosmic stretch factor of 3700 to bring it up to today's size, we get 200 million light-years—the scale of maximum texture, the peak in the roughness spectrum.

So the peak tells us the age of regime change, the time when radiation's early dominance gave way to dark matter. That transition, way back in the Universe's youth, before even the microwave background was made, is imprinted on the patterns of galaxies across the sky. It's an ancient signature for us to read.

Lecture Eighteen
The Dark Age—From Sound to the First Stars

Scope:

In this lecture we follow a very special transition: from the brilliant, smooth, hot young Universe, though a 200-million-year dark age, to the birth of the first stars. It's quite a transformation. After the acoustic era, dark matter lumpiness grows ever-more pronounced as gravity continually slows expansion of the densest regions. After about 100 million years, these regions finally halt, turn around, and collapse to make condensed, dark matter mini-halos. Meanwhile, atomic matter falls into these mini-halos and, unlike dark matter, is able to cool and so collapse further. Finally, after about 200 million years, the collapse is complete, and the first stars are born. These first stars are massive, brilliant beacons that light up the Universe. They live brief lives before exploding in violent supernova explosions that sprinkle the first heavy elements into the surrounding gas.

Outline

I. In the million-year-old Universe, matter was smooth and thin and expanding, but by 200 million years the first stars were forming. How did this extraordinary change occur?

    A. Here's the overall situation.

        1. A smooth, expanding Universe has areas of slightly higher density.

        2. The denser regions expand more slowly, so the roughness grows.

        3. Ultimately, these regions stop expanding and fall back in on themselves.

        4. It's within these collapsed regions that the first stars form.

    B. Making the first stars involves teamwork.

        1. Dark matter gets the roughness to the point that parts can collapse.

        2. Atomic matter's ability to cool allows the collapse to continue to the point that stars can form.

C. This lecture concerns the first half-billion years, which is still unobserved. We'll introduce observations in the next lecture, for the second half-billion years.

II. At the time of the CMB (400,000 years), dark matter's density varies by about 1% from place to place. However, atomic matter is much smoother, with about 0.01% of variations. This makes sense: The CMB patchiness is about 0.01%.

   A. The reason atomic gas is so smooth is that it feels pressure.
      1. Recall, light's brilliance provides the pressure.
      2. Only because the gas is foggy does it feel light's pressure.
      3. Before fog clearing, the gas bounces back and forth as sound waves.

   B. After fog clearing, atomic matter is no longer pushed by light, and it's free to fall toward the denser, dark matter regions.

   C. We can follow the evolution of roughness in both dark matter and atomic matter by plotting roughness, $\Delta\rho/\rho$, against time for a particular wave size.
      1. Initially, both undergo the same super-horizon growth.
      2. When the horizon envelops the wave, atomic matter feels its own pressure, stops getting rougher, and bounces around as sound waves.
      3. Dark matter, however, keeps getting rougher—so by the time the fog clears, it is much rougher than atomic matter.
      4. When the fog clears, atomic matter falls toward the denser areas of dark matter and quickly takes on its rougher form.

   D. If there were no dark matter, atomic matter alone would only have 10% roughness today—a radically different Universe with no stars or galaxies.

III. From fog clearing to a hundred million years, cosmic expansion and red shift make the Universe redder, dimmer, and cooler. After 6 million years, it would be pitch dark to our eyes. After 17 million years, the temperature dips below 0°C, and we enter a long, frigid night called the "dark age."

IV. Around 100 million years, a crucial transition occurs when the densest regions halt their expansion and begin to collapse. Dark and atomic matter undergo rather different kinds of collapse.

    A. Dark matter particles can't collide or release any energy, so their collapse can't continue very far.

        1. After the first chaotic infall, there is a period of adjustment, and the dark matter settles into a fuzzy ball about half its original size. So its overall density is still quite low.

        2. These collapsed objects are called "mini-halos"—they're small versions of today's large galaxy halos.

        3. They are the wombs within which atomic matter can now gather to make stars.

    B. Atomic matter is different in two crucial ways.

        1. Its atoms collide, giving the gas pressure, which slows its collapse.

        2. These atomic collisions also make photons that leave the gas, cooling it. This reduces the pressure, allowing the gas to collapse to a much smaller size, with a much higher density.

        3. It's only from gas at this high density that stars finally form.

    C. This whole process has been simulated using computer programs that include most of the relevant physics.

        1. Typically by about 100 million years, dark matter has collapsed into a few denser mini-halos. When these merge, atomic gas gets pulled in, cools, and settles to the center.

        2. With extraordinary resolution, it's now possible to follow the gas collapse right down to a central region a few light-hours across containing 200 times the Sun's mass.

        3. It is thought that this then collapses to form a single massive star.

    D. These first objects contain 10^6 solar masses of dark matter, 10^5 solar masses of atomic gas, and a single massive star. They are sometimes called "microgalaxies" because they're about a millionth of a present-day galaxy.

    E. The first stars bring "first light" to the Universe and so end the dark age.

V. What were these first stars like?

A. They were made from pure hydrogen and helium, and this results in them being rather different from today's stars, which contain 1%–2% heavier elements.

B. Gas with heavy elements can cool more efficiently, so while today's stars form from gas at 10 K, the first stars form from gas at 250 K.

C. Because this gas has higher pressure, more must gather before it collapses to make a star. So the first stars were very massive compared to today's stars: a few hundred solar masses.

D. Being so massive, they were also incredibly powerful: perhaps a million times brighter than the Sun.

E. With no heavy elements, their surfaces were very hot, around 100,000 K.

    1. They therefore emit strongly in the ultraviolet.

    2. This radiation ionizes the surrounding gas, heating it up and preventing any more stars from forming in that dark matter mini-halo.

    3. So we think these first stars were loners, quite unlike stars today that are born in clusters of hundreds or even thousands.

F. The first stars also had very powerful central furnaces that fused hydrogen into heavy elements at an incredible rate. Within a million years, the fuel is gone and the star dies in a gargantuan supernova explosion.

    1. The debris may contain 100 solar masses of heavy elements that quickly spread over several thousand light-years, "polluting" it.

    2. The newly polluted gas can now cool more efficiently, so that subsequent generations of stars have much lower mass.

G. The period of this lecture is special: It bridges the utterly alien early Universe to our recognizable Universe, with stars and galaxies. Amusingly, matching the cosmic age to a human age, the first stars are born around the 9-month time. We're now ready to follow the infant's growth.

Suggested Reading:

Silk, *The Infinite Cosmos*, chap. 13.

———, *The Big Bang*, chaps. 10, 14.

Jones and Lambourne, *An Introduction to Galaxies and Cosmology*, chap. 6.

Hogan, *The Little Book of the Big Bang*, chap. 7.

Questions to Consider:

1. From ½ million years to 200 million years, the Universe changes from a glowing, expanding, smooth gas to a dark, expanding collection of collapsing clumps with stars at the centers of the densest clumps. How did this remarkable transformation take place?

2. How do dark matter and atomic matter behave differently in the two time periods (a) from the beginning to recombination at 400,000 years and (b) from recombination to the first stars at about 200 million years? In the first period, think about how atomic matter is subject to radiation pressure, while dark matter isn't. For the second period, think about how atomic matter can cool and collapse to form much denser systems, while dark matter can't.

3. What were the first stars like? How were they different from the stars we see in the sky today? Why were they so different? Why were the stars in the second generation so much more like present-day stars?

4. Describe what it would have been like to live through the dark age, from about 5 million years to 200 million years.

Lecture Eighteen—Transcript
The Dark Age—From Sound to the First Stars

Look at my hand, or this pen, or anything around you. The objects in our world are amazingly dense compared to the cosmic average. If my hand was expanded to the average cosmic density, it would fill a sphere that touched the Moon. But back in the young, million-year-old Universe, the matter was incredibly smooth—less than 1% variation in density—and thin, 100 billion billion times less dense than my hand. And also, everywhere was hurtling outwards, a never-ending expansion. So how do we reconcile this apparently hopeless beginning with our current happy ending? Some extraordinary things must have happened to bridge these two very, very different scenes.

Of course, the full story takes 14 billion years, but the most desperate part, the real struggle to make the kind of Universe that we now have, is right at the beginning. Somehow, by just 200 million years, what had been an expanding, thin, hot gas filled with sound waves became a cold, dark Universe sprinkled with brilliant points of light—the first stars. It's really quite an extraordinary and rapid transformation. Actually, if we look at the human parallel that sees 14 billion years as the human lifetime, that 200-million-year period corresponds, appropriately, to the 9-month gestation period. Just as the DNA of the first cell is gradually assembled into a human embryo, so the primordial fluctuations, the cosmic DNA laid down by inflation, grow over this period, transforming beyond all recognition into the first stars.

So the story of how this transformation took place is the topic for this lecture—from sound to the first stars. Here's a diagram that gives you the overall situation. We start with a very smooth, expanding Universe, and dotted about are places of slightly higher density. Now, initially, the expansion is both global (the whole Universe expands) and local (even the denser regions expand). But, and here's the key point from the last lecture, the denser regions expand more slowly, so the roughness grows, and the patchiness becomes more noticeable. Now, there comes a point when these regions stop expanding and fall back in on themselves, so although the global expansion continues, locally, the denser regions have collapsed, and it's within these collapsed regions that the first stars form.

This cartoon is, of course, way too clean. Here's a properly calculated distribution of dark matter at about the time the first stars form. Now, we'll talk about all this filamentary structure in Lecture Twenty-Two, but for now, we're focusing on the densest parts that I've circled. That's where matter has collapsed and where the first stars will form. There's a very important part to the story that I want to stress right at the outset. The road to the first stars involves teamwork, collaboration, if you like, between dark matter and atomic matter. Dark matter first gets the roughness up to the point that the densest parts can collapse, but they don't collapse far enough, and we need atomic matter to take us the rest of the way to make the first stars.

The property that atomic matter has that dark matter doesn't is that it can cool, and this allows it to continue collapsing, right up to the point of making a star. Now, one final introductory point: The overall period we'll be looking at, about the first 0.5 billion years, hasn't yet been seen. It's just too far away. The objects are too faint. So I'm going to postpone talking about observations until the next lecture, when we'll look at the assembly of the first galaxies in the second half of the first billion years.

So let's begin in familiar territory, about 400,000 years after the Big Bang, roughly when the microwave background was made. This diagram shows you just how smooth the Universe was at that time. I've sketched more or less to scale the variation in density from place to place. The orange curve is for dark matter, and it varies by about 1% above and below the average, so the Universe at that time is really quite smooth. But the atomic matter is much smoother. To see its roughness, we need to stretch that 1% way out. Here it is, 100 times less rough than the dark matter, about 1/100 of 1%. That's equivalent to the difference in atmospheric density between your feet and your head. It's really not much. When you think about it, this makes sense. Remember, it's the glowing atomic gas that makes the microwave background, whose patchiness is exceedingly slight, between about 1/100 and 1/1000 of 1%.

Now, the reason the atomic gas is so much smoother than the dark matter is because the atomic gas feels pressure, and the pressure force makes it smooth, just like it makes the air in this room extremely smooth. It's important to remember that the source of pressure is light's enormous brilliance, and the gas only feels this pressure because it's foggy. In fact, we've already encountered exactly how this foggy gas behaves. It bounces in and out of the dark matter clumps. It oscillates as pressure waves, as sound. Now, as long as the gas is foggy and supported by light's pressure, nothing much is going to change, and the atomic gas will remain smooth.

You can probably tell where I'm going with this. What happens when the fog clears, around 400,000 years? From then on, both light and atoms are now independent, and each is now free to move where it wants. Here's our cartoon of the two density patterns. After fog clearing, the atomic matter is now free to move downhill towards wherever the dark matter is denser, away from wherever it's less dense. So the atomic gas quickly gets rougher as it begins to take on dark matter's density pattern. Cosmologists usually show all this using a graph like this, which shows the growth of roughness, $\Delta\rho/\rho$ at the y-axis, against time along the x-axis, for a particular wave size. Both axes are exponential. There are two curves, one for dark matter's roughness, in purple, and the other for atomic matter's roughness, in green.

Now, remember from the last lecture, initially, the horizon is small, so everything undergoes super-horizon growth. Dense regions expand more slowly than less-dense regions, so both lines increase together at early times, on the left. Then when the horizon envelops a dense patch, the atomic matter begins to feel its own pressure, so its roughness stops growing, and instead it bounces in and out of dark matter regions as sound waves, which make these funny ups and downs in the green curve. Meanwhile, dark matter's roughness continues to grow, slowed somewhat by the radiation era, but by the time the fog clears, it is much rougher than atomic matter, just as you saw in the last diagram.

But as soon as the fog clears, the atomic matter is now free to move, so it begins to move towards the denser areas of dark matter, and away from the less-dense areas, rapidly getting rougher. So the green line rapidly increases. By about 15 million years, the 2 lines have almost joined, and the atomic matter is now distributed almost the same as the dark matter. They are in lockstep. Let's pause for a moment to think about what would happen if the Universe didn't contain dark matter. See, in that case, just as before, atomic matter's roughness would get held back until the fog cleared, at which point it would follow this dotted line—normal sub-horizon growth, proportional to cosmic size.

So if it starts at 0.01% roughness at fog clearing, even the factor of 1000 growth from then to now only makes 10% roughness today, not even close to turnaround and collapse. So imagine that—after 14 billion years, we'd still be living in a frigid, dark, sparse Universe, with just 10% density variations from place to place. Not a single star would yet have formed, and probably never would. Now you can see why dark matter is so crucial at getting the Universe on the right path. Because of its independence from light, its roughness can grow at a time that atomic matter is held back by light's pressure. As I've said before, making the first stars requires teamwork.

Now, at this point, I'd like to raise a more general question—what was it like back then? From fog clearing to 100 million years is a new period for us, so what might we have witnessed? Well, expansion, as always, continues to redshift the photons, so the light in the Universe gets redder, dimmer, and cooler. By 6 million years, the radiation is now almost all in the infrared, so for the first time in cosmic history, the entire Universe would be dark to our eyes. Everywhere is pitch black; not a single photon of optical light remains. And for this reason, this marks the beginning of what cosmologists call the "dark age." It's a lovely term, partly because it evokes the dormancy and inaccessibility we associate with that early period of European history.

Now, although the sky is optically dark, the infrared radiation would still kill you. Actually, it would cook you, since the temperature at that time is about the same as your household oven. It's about 450°F. But as time passes, the temperature drops past water's boiling point at around 10 million years, human body temperature at 15 million years, and water's freezing point at 17 million years, and from then on, it gets colder and colder—a long, dark, frigid night falls across the Universe.

But within that darkness, a lot is happening. Gravity is gradually working away to increase the roughness. And by about 100 million years, an exceedingly important transition occurs. Although average density regions keep expanding, the densest regions halt their expansion, turn around, and begin to fall back inwards in a collapse. Here are our expansion curves from the last lecture that show this nicely. The high-density region is already lagging behind, slowed down by its extra mass, but when we extend this diagram to the right, forward in time, you can see how the high-density region behaves just like a little closed Universe. Its expansion halts, turns around, and reverses in a collapse.

Now, this diagram suggests the collapse is an exact reversal of the expansion, ending in a dramatic big crunch—but in reality, that isn't quite what happens. Here's a computer simulation of the kind of collapse we're talking about, starting a little after turnaround. The collapse is rather more chaotic, mainly because the regions are not, in fact, nice neat spheres, so the falling material heads down from slightly different directions and arrives at slightly different times. Also, dark matter particles don't collide with each other, so they all just zoom back out again. Ultimately, after all the chaos of the collapse, the dark matter settles into a big fuzzy ball that's about half its original size, denser at the center and sparser further out.

These collapsed regions are sometimes called "mini-halos" because they're a smaller version of the large dark matter halos we find around today's large galaxies, the kind we met in Lecture Nine. So here's a slightly improved graph of the collapse, with the period of readjustment sketched, along with the final size. Now, a key quality of this kind of collapse is that no energy is released. There's no friction, and no light is radiated away. And cosmologists call this a "dissipationless collapse"—no energy is dissipated. This is extremely important because it means the collapse can't continue very far, and these mini-halos only collapse to roughly half their starting size.

Now, that factor 2 decrease in radius boosts the density by a factor of 8, 2^3, and during the collapse, of course, the rest of the Universe continues to expand, so when the mini-halo finally forms, its density is only about 200 times greater than average, which of course is a significant increase, but it's nowhere near enough to make stars. What we have here are the wombs within which stars can now form. Let's now look at how that happens, how the atomic matter behaves, as it gets caught in the collapse of these dark matter mini-halos.

Well, right off the bat, atomic matter is different in two crucial ways. First, its atoms collide with each other, and this gives the gas pressure, which of course tends to slow the collapse. But those very same atomic collisions can also create photons that leave the system. They might be infrared or light or radio photons—it doesn't matter. They remove some of the energy of the gas, so because the gas radiates, it can cool. Its pressure drops, so gravity can pull it down to a much smaller size, where it reaches a much higher density. Cosmologists call this kind of collapse "collisional and dissipative." The atoms collide to make pressure, and they dissipate energy, and these two qualities are crucially different from dark matter, and they allow the atomic matter to collapse much further down and reach the kind of densities needed to form stars.

Now, ultimately, what gives atomic matter this advantage is that it contains tiny charges, electric charges—electrons and protons. And so when atoms get close, they deflect. They collide, and the jostled charges make photons. See, dark matter particles have no charge. They neither collide nor make photons. Remember, they're dark. So let's now look at some recent attempts to model this combined dark matter–atomic matter formation of the first stars.

By now, you should be getting used to the way in which cosmologists work. In addition to making observations and developing physical theories, they also try to calculate what happens on a computer as accurately as possible. Here's one of those computer simulations. As you'd expect, both atomic and dark matter start out very smooth, but by about 100 million years, the dark matter—along the bottom—has collapsed into a few denser mini-halos that form at the intersection of a web of lower density filaments. It's in these denser mini-halos that the first stars will form. Now, above, you can see how the gas basically follows the dark matter. Initially, its pressure prevents it from falling into the smaller dark matter regions, so it's fuzzier, but as these smaller regions merge and grow, they are massive enough—perhaps a million times the mass of the Sun—to pull in the gas, where its ability to cool causes it to collapse into a very small volume, utterly invisible on this plot.

Now, this movie shows a zoom into that central clump, starting from a region a few thousand light-years across and ending with a region a few light-hours across. It's equivalent to zooming into a city to see a single ant sitting on your finger. That final region is roughly the size of the solar system and contains about 200 solar masses of gas, moving inwards. Now, at that point, the computational methods hit their limit, but it seems fairly certain that those 200 solar masses will collapse down into a single massive star. These and similar calculations all suggest a single massive star forms at the center of a dark matter mini-halo. The halos are a few thousand light-years in size. They contain about 1 million solar masses of dark matter, 100,000 solar masses of atomic gas, and a single star at the center.

These first objects are sometimes called "microgalaxies" because they're about a millionth of the mass of the present-day Milky Way–type galaxy. These microgalaxies are dotted around the young Universe wherever there were peaks in the original lumpy distribution of dark matter. So a typical separation between two stars might be 10,000 light-years—about the distance between the Sun and the center of our galaxy. These first stars provide what is called "first light." The Universe ended its 100-million-year, dark, cold night in a wonderful new dawn as the light from this new population of brilliant stars flooded out into space.

Let's now look at these first stars in more detail. For example, were they like the Sun or other stars in our sky? The answer is "probably not." We think they were very unusual compared to today's stars, primarily because their composition was different. They were made from essentially pure hydrogen and helium, with no heavier elements. Now, in today's Universe, atomic matter contains 1% or 2% heavier elements, things like oxygen, carbon, or iron. And even this small fraction significantly changes how stars form and how they live out their lives. The crucial property that heavy elements contribute is an increased ability of the gas to cool. When atoms of heavier elements collide, their more complex electron orbits are jostled more easily, which makes photons that can carry away energy, cooling the gas.

So in today's Universe, a star-forming gas cloud is a very frigid 10 K, but in the young Universe, a pure hydrogen and helium gas cloud couldn't cool below about 250 K. Now, why is this relevant? It's because hotter gas has higher pressure, and this resists gravitational collapse, so more mass must gather before a final runaway collapse can make a star. Bottom line is stars born from higher temperature gas are more massive. In the case of the first stars, they're thought to be a few hundred times more massive than the Sun.

Now, such massive stars really don't exist in today's Universe. Most stars are around the mass of the Sun or less, and a few can reach a maximum of about 60 solar masses. These first stars were truly unusual by today's standards. So what were they like? Now, you may be wondering how we can know anything about these first stars, but computer modeling of stars is now a very mature subject. One can actually follow the lives of these massive stars in considerable detail. For example, massive stars are incredibly powerful for the simple reason that their great weight compresses and heats their central furnace, which burns extremely vigorously. These powerful furnaces make the first stars shine with incredible brilliance, perhaps a million times brighter than the Sun. So if we were orbiting one of these first stars, it would be as big as your fist in the sky, and the entire biosphere would burst into flames in less than a second.

The absence of heavy elements also makes the star's surface very hot, perhaps 100,000 K—20 times hotter than most stars today. And with such hot surfaces, much of their radiation falls in the ultraviolet part of the spectrum. So, to our eyes, these stars would appear bluish white. Now, ultraviolet light is extremely destructive to atoms. It can knock the electrons right off them. It ionizes atoms. So these first stars ionize their surrounding gas and heat it up, which has the effect of preventing any more stars forming in that dark matter mini-halo. It's a bit like a cuckoo, hatching out in some unsuspecting bird's nest. By pushing all the other eggs out of the nest, it ensures it's the only surviving chick. So we think these first stars are loners. Born individually, they live out their brief lives in relative isolation. Again, this is quite unusual, compared to stars in today's Universe, which are born hundreds or even thousands at a time from the same cloud of gas.

Now, remember, the primary reason these first stars are so unusual is because of their pristine composition—just pure hydrogen and helium, with no heavier elements. So let's now look at how that condition began to change, because it did so as soon as the first stars began to die. Perhaps the most important fact about all stars is how they live, how they make the energy that makes them shine. Deep inside the star, nuclear reactions gradually convert hydrogen into heavier elements, things like carbon, oxygen, and iron. Exactly which elements are made and how fast they're made depends primarily on the mass of the star. Now, these very heavyweight stars have extremely powerful and hot furnaces that burn through their hydrogen fuel at an incredible rate, and within just a million years, the fuel is all but gone, and the star begins to die.

The way in which these massive stars die is one of Nature's most spectacular sights. It's a gargantuan supernova explosion that completely disrupts the star and hurls its guts, including all those new heavy elements, out into space. In the case of these first stars, it's thought that the debris may contain 100 solar masses of heavy elements that quickly spread over several thousand light-years, polluting, in a sense, the environment. This pollution turns out to be very significant. If those 100 solar masses of heavy elements mix into the surrounding 100,000 solar masses of pristine gas, we end up with about 0.1% pollution. Now, this may not sound very much, but it is, both in human and cosmic terms.

On a very bad day, the fraction of smog pollutants in a major city's air is about 0.0001%. If the smog was 0.1%, you could barely see your hand in front of your face. Now, cosmically, this degree of pollution completely changes the ability of the gas to cool. This is very important because the next generation of stars to form from this polluted gas will do so from much colder gas, and so all future generations of stars are of much lower mass, much more like the Sun.

The period we've looked at in this lecture is really quite special. It acts as a bridge between the utterly alien realm of the smooth, hot, early Universe and the old, cold Universe of today with its familiar stars and galaxies. So let's just review how this remarkable transformation took place, with these images. The patches on the microwave background remind us that the hot, early fireball contained extremely slight variations in density from place to place. After fog clearing, the fireball faded, and with ever-deeper darkness, the roughness of both dark matter and atomic gas slowly grew larger and larger.

After about 200 million years, the densest parts halt their expansion, turn around and collapse, to form mini-halos of dark matter. And within these mini-halos, atomic gas cools and continues to collapse, finally leading to a dense core. With only poor cooling available, this warm core undergoes runaway collapse to form a very massive star. These first stars live brief but brilliant lives, which light up the whole Universe for the first time since the original fireball. The stars quickly die in powerful supernovae that spew freshly made heavy elements out into the surrounding gas, forever changing the way in which future generations of stars will form.

Returning to our human metaphor, we see these first 200 million years as the dark period of gestation that slowly assembles the first structures. It ends, of course, dramatically with the birth of these first stars, which marks the emergence of the kind of Universe we now find ourselves in. As we'll see in the next lecture, the newborn infant has emerged, kicking and screaming, and the fireworks have just begun.

Lecture Nineteen
Infant Galaxies

Scope:

Following the last lecture's story of the first stars, we continue with the first galaxies. A key idea is "hierarchical assembly": Stars form in groups that merge to form star clusters, which form in groups that merge to form infant galaxies. This early phase of collisions and mergers occurred amazingly rapidly because the Universe was so much denser than it is today. Using the Hubble Space Telescope, it's now possible to study galaxies back to 1 billion years, where we find small, chaotic, interacting, blobby systems. It takes time—and calm conditions—to construct today's majestic spirals. Closely related to the buildup of galaxies is the buildup of their star populations. The history of star birth shows a "baby boom" around 1–3 billion years, followed by a steady decline to today's relatively infertile Universe. Future instruments should reach the first half-billion years and hopefully see the first stars.

Outline

I.   The last lecture visited the first half-billion years, ending with the first stars. This lecture visits the next billion years and the birth of "infant" galaxies.

   A. It's a spectacularly active period of assembly.

   B. It's also accessible to our best telescopes, so the story is based firmly on what we witness.

   C. Matching cosmic and human lifetimes, we're studying children up to age 10. Just as this period lays the foundations for adult life, the remaining 11 billion years of galaxy maturation are a natural extension of the first 2 billion years.

   D. Two key ideas will be important to us.
   1. Smaller things merge to make bigger things, in a hierarchy of assembly.
   2. This assembly was very rapid compared to today's sedate Universe.

II. Let's look at these ideas of hierarchical assembly and rapid initial growth.

    A. The first regions to collapse have the highest density. Where are these?

    B. From Lecture Seventeen, recall the density pattern can be thoughts of as the sum of long, medium, and short "waves" of density. So the densest regions are where all the peaks line up: small on top of medium on top of long waves.

    C. The first peaks to collapse make stars. But these sit on top of larger peaks, so they're born in groups that then collapse to make star clusters. But these in turn form in groups that collapse to make galaxies.

    D. This ongoing assembly of larger systems from smaller ones is called "hierarchical merging" and is sometimes illustrated using a "merger tree": twigs connect to branches that connect to a trunk.

    E. Computer simulations nicely show this process in action.

        1. They typically follow the motion of millions of dark matter particles, each one pulled by all the others, so it's a complex, self-interacting system.

        2. The movies usually adopt a "comoving" viewpoint that doesn't show cosmic expansion. This greatly clarifies what's going on.

        3. Starting from a very smooth distribution, the first structures form very quickly and then continue to grow as more matter rains down.

    F. The reason the initial structure growth is so rapid is fairly straightforward.

        1. The Universe was smaller and denser, and dense things collapse quickly because their gravity is stronger and everything is closer.

        2. In general, when a region has halted its expansion, its collapse time is equal to the cosmic age at that time. For example, turnaround at 200 Myr means collapse by 400 Myr, which is fast by today's standards.

III. By looking far away, long ago, the Hubble Space Telescope can study these first systems. It's difficult work: They appear the size of a football stadium on the Moon, glowing with the brightness of a single candle.

   A. The Hubble Ultra Deep Field (HUDF) is a wonderful example: Not much bigger than a pinhead at arm's length, it contains 10,000 faint galaxies (suggesting 100 billion cover the sky), most around 1 arc second in size.

   B. The HUDF is like a long tunnel back in time, so it's crucial to measure distances to the galaxies, to get their ages. Standard methods don't work for these faint galaxies, so a different method is used.

   1. The HUDF is really 4 images taken through 4 differently colored filters that let through a narrow range of colors: blue, green (visual), red, and infrared, which we'll call B, V, R, I.

   2. Young galaxy spectra have a sharp drop in flux near 100 nanometers (the ultraviolet).

   3. Because more distant galaxies have higher red shift, this blocked region moves to longer wavelengths, through the 4 filters. Once a filter is passed, it receives no light and is said to have "dropped out."

   4. For example, a galaxy with redshift stretch factor 5 is visible in the V, R, I images but not the B image. This is called a "B dropout" galaxy.

   5. B, V, and R dropout galaxies have redshift stretch factors of 5, 6, and 7, with ages of 1.6, 1.2, and 0.9 billion years after the Big Bang.

IV. Let's compare these young galaxies to today's galaxies—it's a bit like comparing children with adults.

   A. They're much smaller, typically only 10,000 light-years across. Plotting galaxy size versus age clearly shows them getting bigger over time. This nicely matches our theory of hierarchical assembly: Smaller things make bigger things over time.

   B. This is also clear from their appearance. Many are closely interacting.

C. There are no pretty spiral galaxies; they all appear chaotic and blobby. In fact, the first modern-looking spiral galaxies don't appear until 6 or 7 billion years; it takes time and calm conditions to build them.

D. We think present-day "dwarf irregular" galaxies may be a good match to these first infant galaxies.

V. Because galaxies are made of stars, in studying the birth of galaxies we also need to study the birth of their stars. What's the star birth history for the Universe?

A. There are many methods that astronomers have developed to measure the birth rate of stars, and it's possible to apply some of these methods to distant galaxies. Piero Madau first did this in 1996, so a graph of star birth rate from the Big Bang to today is sometimes called the "Madau plot."
  1. It has a rapid rise in the first billion years and reaches a peak between 1 and 3 billion years: a baby boom!
  2. There is a steady decline after that, so today's birth rate is only 5% of what it was at the peak.

B. Why was there a stellar baby boom? Because galaxy collisions trigger the birth of stars, so when the pace of collisions was high, so was the star birth rate.

C. This early period would have been spectacular to witness: galaxies colliding, brilliant young stars, supernova explosions. Quite different from today's relatively tranquil Universe.

VI. It should soon be possible to see yet earlier times.

A. One can push the "dropout" method further into the red using near-infrared filters. A new camera on HST, due to be installed in late 2008, should be able to do that and find galaxies out to red shifts near 11, in the first half-billion years.

B. Our best hope is the James Webb Space Telescope (JWST), due to be launched in 2013. It's a large infrared-optimized telescope that should see the first stars.

Suggested Reading:

Silk, *The Big Bang*, chaps. 10–11.

Keel, *The Road to Galaxy Formation*, chaps. 5, 10.

Hogan, *The Little Book of the Big Bang*, chap. 7.

Questions to Consider:

1. How are the first galaxies assembled? Starting from the first stars, describe the early stages in the hierarchy of assembly. Why are the first stars born in groups? Think about where the densest parts are—they are where the peaks of short, medium, and long density waves all line up.

2. Why do the early stages of galaxy assembly occur so rapidly compared to today's much slower pace of galaxy development?

3. In the deepest Hubble images, how do astronomers decide which galaxies are near and which are far? What's the current limit for the most distant galaxies seen in these images—how long after the Big Bang can we see?

4. In what way are infant galaxies different from today's galaxies? Roughly when did the first familiar-looking, pretty spiral galaxies appear?

5. Describe the cosmic history of star birth. Why were stars born so rapidly in the first 3 billion years? Why has there been a steady decline since then?

Lecture Nineteen—Transcript
Infant Galaxies

This is the third lecture in our group of six lectures on the growth and evolution of structures, things like stars and galaxies. So far, we've done well. We've seen how minuscule ripples, emerging from the Big Bang's inflationary launch, steadily grow more and more prominent, and in the last lecture, we saw how the densest parts slowly halted their expansion and collapsed to form the first stars.

Here's a 14-billion-year timeline, from the Big Bang to now, with a blowup of the first 2 billion years. So the last lecture's story took us through the dark age, the collapse of mini-halos, and to first light, with the birth of the first stars, and all this took about 0.5 billion years. We still have a long way to go to get to our current Universe. So what happened next? Well, roughly speaking, the next couple of billion years saw a rapid and spectacular assembly of what one might call "infant galaxies," rather chaotic collections of stars and gas that were much smaller than today's huge, majestic galaxies. They're sort of little galaxy building blocks.

Now, for reasons we'll get to, this is a fabulously active period in the life of the Universe. It's packed with galaxy collisions, star birth, and supernova death. And the good news is our best telescopes can now see out to this region, so the story is based firmly on what we see. This is the theme for the lecture—the wild exuberance of youth. Back to our human lifetime metaphor, we'll be looking at roughly the first 10 years of life. Now, for humans, this is a crucial time. By 10, most of the foundations for adult life are in place. So, too, with the Universe—and the remaining 11 billion years of galaxy growth and maturation, which we'll tackle in the next lecture, will be a natural extension of these first 2 billion years.

Let's start with two very simple ideas. The first is that little things merge to make bigger things, which merge to make even bigger things, and so on, in a hierarchy of assembly. Now, the second idea is that relative to today's sedate Universe, this early assembly process went on very rapidly. So let's look at the first idea, the hierarchy of assembly. This diagram illustrates the basic process.

Remember from Lecture Seventeen that we can think of the actual complex pattern of density variations from place to place as the sum of short, medium, and long waves, just added together. Now, the regions that are going to collapse first are those of highest density. Let's see where they are. They're where peaks sit on top of peaks that sit on top of peaks. The smallest peaks will collapse first to form stars, but because these peaks sit on top of larger peaks, then the stars are born in groups, which collapse to form star clusters, and these, in turn, are formed in groups, which collapse to form galaxies.

Now of course, in reality, there aren't these discrete stages, and instead there's a continuous buildup of ever-larger systems, with smaller ones forming all the time. Cosmologists call this continuous buildup of larger systems from smaller ones "hierarchical merging," and it's a key idea in the long road of assembly, from the first lone stars in the newborn Universe to the giant clusters of galaxies in today's old Universe. All the way along, larger structures are formed from the successive merger of smaller structures. You can illustrate this using what's called a "merger tree," which shows how many smaller structures gradually come together to build larger ones in the growing hierarchy. I'm afraid I plotted this one with time going upwards to match my earlier diagram, so I've stood the tree on its head.

Let's now bring these cartoon diagrams to life by showing a computer simulation of this growth of structures. This first movie gives the big picture, a giant cube that ultimately expands to become 150 million light-years on a side that contains thousands of galaxies. A supercomputer is used to calculate the motion of 2 million dark matter particles, each one following the path that gravity dictates, and the gravity, of course, comes from all the other dark matter particles, so it's a complex, self-interacting system. Now, I must warn you the camera viewpoint in all of these movies expands with the Universe, so you won't see any of the cosmic expansion itself. This is common practice in cosmology and is called a "comoving" viewpoint. It helps you focus on the structure formation and not get distracted by the expansion.

This particular calculation starts about 0.1 billion years after the Big Bang, when everything was 30 times closer together. Notice how smooth the distribution of matter was. You can hardly see the tiny fluctuations that are spread throughout the region. The movie will last just 12 seconds, and spans all of cosmic history, so it's about a billion years per second. It's probably the fastest action replay you've ever seen. Okay, off we go. Notice how extremely quickly the texture emerges, and as time passes, its form gets more and more defined. Matter appears to be moving away from the voids and towards the filaments and clusters. We'll have a look at this filamentary structure more in Lecture Twenty-Two, but for this lecture, let's zoom into one of the peaked regions, about 1/10 of this larger size, and watch the formation of a relatively small group of galaxies. Notice how quickly the first structures form, and then more and more material piles into the region, merging with what's already there.

Now, the theme of this lecture is the first 2 seconds of that movie, when the first structures formed amazingly quickly out of the smooth initial background. Why was that so fast? The main reason is fairly simple. The Universe was smaller and denser back then, and simply put, dense things collapse quickly, in part because gravity's stronger, so the infall speeds are higher, and in part because everything is simply closer together, and it doesn't take long for two neighboring regions to collide and merge. In fact, there is an interesting, simple rule of thumb. When a region is just ready to collapse, the collapse duration is roughly the age of the Universe at that time. So 0.1 billion years after the Big Bang, it takes about 0.1 billion years for regions to collapse, while 2 billion years after the Big Bang, it takes 2 billion years for a region to collapse. This is pretty obvious when you consider this sketch from the last lecture. The expansion and contraction of a higher-density region is roughly symmetrical. The downward infall takes about as long as the expansion up to turnaround. See, stated like that, it's now clear why there is such a frenzy of collapse and merger at the beginning. The Universe is young and relatively dense. The collapse time is short, so the overall pace of change is fast, relative to today's much more sedate pace.

Okay, now that you have the theoretical framework down, it's time to shift gears and look at real observations of this youthful, billion-year-old Universe. And I'm glad to say that in the last decade or so, we have now been able to gather increasingly detailed observations of these very early times. By now, you should realize this means looking extremely far away, to regions whose light has been traveling for 12 or 13 billion years. Only then can you witness the 1-billion-year-old Universe. An absolutely key instrument in this endeavor has been the Hubble Space Telescope, with its exquisitely sharp vision. At these enormous distances, the subgalactic blobs are exceedingly tiny and faint. Typically, they appear about as big as a football stadium on the Moon, glowing with the brightness of a single candle. Not surprisingly, then, the strategy has been to look long and hard at small patches of the sky, to go as deep and faint as possible.

Here, for example, are the approximate footprints of four of these deep-survey fields, compared to the full Moon, which itself seems pretty small put next to the Big Dipper. They go by various acronyms and are designed for slightly different projects, but perhaps the most famous are the Hubble Deep Field, which was completed in 1995, and its more sensitive successor, the Hubble Ultra Deep Field, which was completed in 2004. Each of these covers a tiny region of the sky, about the size of a pinhead held at arm's length.

Here's a couple of comparisons to show you how advanced these deep surveys have become. The first compares images taken from the ground and from space, where the absence of atmospheric turbulence makes a stunning improvement, especially to images of the smallest galaxies, which are otherwise just smeared out into smudges by what astronomers call "atmospheric seeing." But to see the faintest galaxies, one also needs very long exposures. Here, the 400 orbits of exposure for the Ultra Deep Field clearly show galaxies that are barely seen in the shallower, 13-orbit exposures of the GOODS survey.

So let's take a look at this famous Hubble Ultra Deep Field. Here's the entire image, which only really comes to life when you zoom into its roughly 6000 by 6000 pixel frame. Actually, if you go to the Hubble website, you can roam and zoom across the entire image. To give you an idea of the quality, here are blowups of three small regions containing large galaxies at distances of 4, 6, and 9 billion light-years. But a more typical patch contains dozens of tiny smudges, with a variety of shapes and colors, and most are smaller than 1 arc second in size, equivalent to the size of the head of somebody standing 30 miles away.

Ultimately, after close inspection, about 10,000 separate galaxy images are found. Of course, there's nothing special about this particular piece of sky. Remember from Lecture Three, the Universe is isotropic on these large scales, so we could have looked in any direction, and we'd find a similar number of faint galaxies, totaling then 100 billion across the entire sky. Now, when you look at this image, you must remember that you are seeing down an extremely long, narrow tunnel, and these galaxies are scattered along that tunnel at different distances. So to use this image to study galaxy evolution, it's very important to measure distances to the galaxies, because only then can you separate infants from children, from adolescents, from adults.

Now, in Lecture Six, I described several ways to measure distances to galaxies, using Cepheid variable stars or the Tully-Fisher method or supernovae and so on. But those methods don't work on these galaxies—they're way too faint and far away. In fact, most are so faint that it's not even possible to take a spectrum to measure a red shift. We really need a different method that works just from the images alone. Well, fortunately, Nature is kind to us, and there is another way. It's not as accurate as taking a spectrum and measuring a red shift, but it amounts to the same thing. You see, the Hubble data set isn't just a single image. There are actually 4 images taken, through 4 different colored filters. A yellow filter, for example, blocks all colors except yellow.

Hubble has many filters it can place in front of its camera, like the one shown here, and every galaxy looks a little different through each filter. The nice, pretty, colored images we're so used to seeing are actually made by combining these various filtered images, using a computer to specify each color. The Ultra Deep Field used 4 colored filters, and this graph shows the colors they let through. They're transparent over a particular range of wavelengths, which gives the filter its name: blue, visual, red, and infrared—or B, V, R, I for short. Now, actually, for those in the know, these aren't quite the names used by astronomers.

So here's the critical part. Let's look at the spectrum of a galaxy, overlaid on these filter curves. Notice that the light spans a very wide range of colors, but down in the ultraviolet, there is a pretty sharp decline near 100 nanometers, which is caused by hydrogen gas in the stars within the galaxy, which blocks light with wavelengths less than about 100 nanometers. Now, watch what happens as we look at more and more distant galaxies. Remember, their spectra are more and more redshifted, so this galaxy spectrum moves to the right. All wavelengths are stretched by the cosmic stretch factor. Now, I've labeled the stretch factor, the distance, and the age since the Big Bang. Remember, at higher red shifts, we're looking further away, and so further back in time.

Now, when you get to a stretch factor of about 4, the blocked region has moved into the B filter, and by about 5, the B filter gets essentially no light, while the V, R, and I filters still get light. So here's an example of this situation. You can see the galaxy in the V, R, and I images, but it's completely gone in the B image. We call this a "B dropout," and we know this galaxy must be extremely distant, and therefore very young. Here are some other B dropouts at stretch factor 5, which corresponds to about 1.6 billion years after the Big Bang, or about 9-year-olds on the human lifespan.

But the method doesn't stop with B dropouts. At even higher red shifts, the blocked region moves through the V filter, and we have V dropouts, with an age of about 1.2 billion years, or 7-year-olds. And finally, at a redshift stretch factor of 7, even the R filter drops out, and we have the R dropouts at about 0.9 billion years, or 5-year-olds. So let's just summarize these three groups on our timeline. You can see that the B, V, and R dropouts really push back into the first billion years, close to, but not quite to, the first stars.

Let's look more carefully at these young galaxies. I know they don't look too impressive, but there's still a great deal that can be learned about them from these images. Overall, we'll be comparing them to today's galaxies. We're comparing children with adults. So right off the bat, just like human children, these young galaxies are quite small compared to today's adult galaxies. Here are some of the brightest B dropouts, and for comparison, plotted at the same physical scale is a normal galaxy from today. The difference is really quite remarkable. Essentially, there were no big galaxies back then—everything was of small, subgalactic size. This is, of course, just what we expected from the theory and the simulations.

The first objects to collapse are the small, high-density peaks, and larger objects only build up later when the smaller ones merge during the turnaround and collapse of larger regions. In fact, this graph shows exactly this effect. You can see how the average galaxy size was about 10,000 light-years for the first 2 billion years, but then it began to increase, presumably as galaxies grew by the merging of these smaller galaxies. Actually, there's strong circumstantial evidence for growth by merging because a very high fraction of these faint young galaxies seem to be gravitationally interacting in close pairs and small groups. Now, of course, our view is essentially frozen in time, but the scene is extremely similar to the simulations I showed, in which there's a flurry of galaxy-galaxy collisions and merging going on.

Another feature of these young galaxies is that their shapes are much more chaotic than galaxies in today's Universe. In this collection, there isn't a single pretty spiral or smooth elliptical—they're just blobs. Now, perhaps that isn't too surprising, given all the collisions, but even in the apparently isolated galaxies, there's none of the regular symmetric forms that we find in today's large galaxies. In fact, coming forward in time, it takes about 6 or 7 billion years before we see the first nicely patterned, modern-looking galaxies. This little sequence of galaxies, plotted at the same physical scale, shows what's going on. We start with small, amorphous galaxies, pairs and groups, and only the last galaxy, at an age of about 9 billion years, has a nice spiral appearance.

Now, although we can't see too much detail in the most distant galaxies, we think they looked like a particular kind of galaxy in today's Universe called a "dwarf irregular" galaxy. Here's an example, also photographed by the Hubble telescope, and I think it gives you a better idea of what these young galaxies might have looked like. Now, although I've been talking about the growth and assembly of galaxies, there's a more essential transformation going on here. It's the birth of stars, which themselves of course make galaxies. It's a bit like discussing the growth of nations. You need to think about the growth in population, and that in turn depends on the birth rate. For example, here's the birth rate of the United States from 1930 to today. It's a crucial graph for any policymaker, not least because you can clearly see the postwar baby boom, with its implications for an aging population in the next few decades.

Well, the Universe also has a population of stars, and that population has grown as the billions of years have gone by. So, not surprisingly, one of the key aims of cosmology is to understand this population of stars. An important aspect of that is the star birth history of the Universe, so let's look at that now. You may be wondering how we can measure the birth rate of stars, but although I don't have time to explain how, it turns out that there are several telltale signs of stars being born. And over the past 30 years, astronomers have carefully studied those signs and learned how to use them to measure the birth rate of stars.

One of the first people to try to construct the entire cosmic star birth history was Piero Madau in 1996, and since then, the methods have gradually improved, although there are still some significant uncertainties. So here, then, is a slightly simplified version of what is often called a "Madau plot," showing the star birth rate from the Big Bang to today. As you can see, there is in fact a baby boom. It happened very early on. After a very rapid rise in the first billion years, the birth rate maxed out somewhere between 1 and 3 billion years, after which it's been declining ever since, so that today's star birth rate is down around 1/20 of what it was during the baby boom period.

Now, just as with human baby booms in the postwar years, it's important to know why there was a stellar baby boom in the early Universe. The reason is thought to be fairly straightforward. When galaxies collide and merge, stars are born in great numbers, for reasons, actually, we'll get to in the next lecture. And because the pace of galaxy collisions was so high in the early Universe, so the star birth rate was also very high. Now, having shown you the history of star birth, there is of course a related quantity—the growth in the total star population. As you might expect, the total number of stars steadily increases as the Universe gets older. Basically, stars are being born faster than they're dying, and so there's a net buildup, and here we are in a Universe filled with stars.

Now, it's important that we don't let this brief, billion-year period of frenzied star births slip by without pausing to appreciate its spectacular nature. Any sentient life forms watching the cosmic scene at that time would be vastly more impressed than they'd be with today's subdued and empty scene. It probably looked something like this artist's impression, showing a Universe strewn with dwarf galaxies. Now, don't forget, everything was 5 times closer, so a high fraction of these dwarf galaxies are colliding and merging. Along with all the star birth goes star death, when the most massive stars explode as supernovae. So the galaxies are studded with brilliant stars, laced with the glowing pink gas of star formation, and perforated by expanding cavities driven out by supernova explosions. It's really a fabulous spectacle.

Actually, back to our timeline, we can now add a history of starlight. Brightness grows quickly after the dark age and the first stars, peaks for a few billion years, and then slowly fades towards today's much dimmer Universe. Now, you may be wondering whether it's possible to see yet earlier times. After all, the most distant R dropouts are near 0.7 billion years, and the first stars were at 0.2 billion years, so what about that period in between? Well, as of now—that's 2008—we haven't quite got there, but I'm going to end this lecture looking at what might soon be possible.

The most obvious strategy is to push the dropout technique to higher red shifts. Here's the filter system we met before, but with 2 more filters added in the near infrared, called "J" and "H." Hubble used a different camera, an infrared camera called "NICMOS," to take long-exposure images of the Ultra Deep Field using these filters. Back on our timeline, the I filter drops out at a stretch factor of about 9, while the J filter drops out around 12, very close to the first stars—which of course is a holy grail for cosmologists. In human terms, these 2 filters isolate 3- and 2-year-olds, so they're like galaxy toddlers. So what have the results been? I'm afraid, so far, not too much—4 possible I dropouts, and 1 possible J dropout.

But that's only the current situation. Let's finally take a quick look at what new technology will soon make possible. In late 2008, the Hubble Telescope will have a new camera installed, called "Wide Field Camera 3," which has a significantly improved sensitivity in the near infrared. As these graphs show, the old camera fails to see galaxies beyond a redshift stretch factor of 8, but the new camera is expected to find galaxies out to stretch factors of 11, well into the first 0.5 billion years—almost, but probably not quite, to the first stars.

But, of course, there's no substitute for a specially designed, infrared-optimized large space telescope, and that's the next hope for this area of research. I mentioned it before in Lecture Eight. It's the James Webb Space Telescope, due to be launched in 2013. Here you can see the comparison of mirror sizes, and here's a simulation of a Deep Field. If you were impressed by the Hubble Ultra Deep Field, the JWST Deep Fields will be stunning by comparison, just littered with galaxies, right back to the first stars.

This is, of course, what we can expect in the next two decades. But hopefully, 100 years from now, we should understand the birth and development of galaxies, just as we now understand the birth and development of humans.

Lecture Twenty
From Child to Maturity—Galaxy Evolution

Scope:

How do galaxies mature from small, blobby infants to the majestic spiral and elliptical galaxies we see today? It's a 12-billion-year story that has a number of themes. Most important are galaxy-galaxy interactions, of which there are many kinds. We'll look at gentle flybys that induce S-shaped arms, galactic cannibalism in which a big galaxy slowly shreds and "eats" a smaller galaxy, and head-on collisions between spiral galaxies that ultimately make an elliptical galaxy. Gas is an important player in galaxy collisions: It falls into galaxy nuclei, where it forms new stars at a prodigious rate. Galaxy anatomy is constructed: bulges and halos via dramatic collapse and collisions, and disks via gentle infall of surrounding gas. Environment also plays an important role. Disks are fragile and can't survive within a galaxy cluster. We end by watching a simulation of the future collision between our own galaxy and our neighbor galaxy, M31.

Outline

I. The last lecture ended around 2 billion years. Let's pick up the story and follow those blobby infant galaxies through 12 billion years of maturation. How did they turn into today's majestic galaxies? There are a number of processes at work.

II. The first major process is galaxy-galaxy collisions. These can range greatly, from flybys, to minor collisions, to head-on impact, to multigalaxy pileups!

    A. Our understanding of galaxy collisions has been a close collaboration between observations and computer simulations.

    B. The first simulation in the early 1970s followed what happens to the stars orbiting one galaxy when a smaller one passes by.

        1. The gravitational pull by the smaller one disturbs the orbits of the stars in the larger one.

        2. The resulting pattern has 2 large "tidal arms" making an S shape.

      3. These S shapes are often seen in real interacting galaxies.

  C. Simulations are now much larger and routinely include 100 million particles that are shared between bulge, disk, and dark matter halo.

  D. Here's a simulation of a relatively gentle process called "galactic cannibalism."

      1. A small galaxy orbits a larger one.

      2. The large galaxy's gravity slowly shreds the small one: Stars get pulled off and form long "tidal streams" that orbit the large galaxy.

      3. Ultimately, the small galaxy may be completely destroyed.

      4. It is thought that all major galaxies "eat" smaller ones this way, including our own Milky Way galaxy.

  E. A simulation of a head-on collision between 2 large spiral galaxies is much more dramatic!

      1. Some spiral structure forms as they approach, but at closest passage the disks are completely destroyed and huge tidal arms flung out.

      2. During the fly-through, no stars collide—they're just too small compared to their separation.

      3. Ultimately, the 2 galaxies merge to make a single galaxy that looks just like an elliptical galaxy. It seems spiral galaxies can change into elliptical galaxies when they merge!

      4. As before, the zoo of real galaxies includes many that look like the simulation, in all its stages.

III. What happens to the thin gas between the stars during these collisions?

  A. Calculating how gas behaves is much more complicated than either stars or dark matter. Only since the 1990s have computers been powerful enough to include it in the simulations.

  B. Unlike the stars, the gas from each galaxy collides at high speed, which compresses and heats it.

      1. Some hot gas expands out into the halo.

      2. Some gas cools and goes to the galaxy center.

C. An important question is whether stars form from the gas during galaxy collisions. The answer is "yes," in huge numbers!

  1. Gas compression can help trigger star birth.

  2. The gas that goes to the center is also prone to forming stars.

D. Observationally, regions forming many stars emit large amounts of infrared radiation.

  1. Interacting galaxies are like beacons to infrared-sensitive telescopes.

  2. Typically, 100 stars are born per year, compared to 0.1 per year in a normal galaxy, like the Milky Way.

  3. These are called "starburst galaxies" because they have such high star formation rates.

E. We can now understand better the Madau plot from the last lecture.

  1. Recall, it shows the history of star birth: a steep rise, peak, and steady decline to today.

  2. This is also the history of galaxy collisions. The Hubble telescope finds almost all galaxies colliding in the young Universe, a steady decline, and 2% colliding today.

  3. The decline in collisions is because cosmic expansion separates the galaxies, and they've already merged, so fewer remain.

IV. Let's now look at the origin of modern-day galaxy anatomy. How do bulges, halos, and disks form?

  A. We've already met how halos and bulges form: They're assembled from the initial collapse and during ongoing collisions and mergers.

  B. Galaxy disks, on the other hand, grow during the relatively quiet times between mergers and involve a gentler acquisition of gas that falls in from the surroundings.

    1. Simulations that include gas show how it settles toward the center of a dark matter halo and forms a rotating thin disk.

    2. Stars form within this disk and build up a population. That's the current situation for the Milky Way: We've not had a major merger for 10 billion years, and gas and stars (including our Sun) have slowly built up.

       3. The stars within galaxy disks gently pull on each other and form wonderful patterns of spiral arms, rings, and bars.

V. Galaxies can also be subject to environmental effects.

   A. Galaxy disks, for example, are fragile and can be destroyed easily. This is obvious when you look at where they live.

       1. Disk galaxies are found mainly in empty regions and rarely in rich galaxy clusters.

       2. In fact, there is a steady decline in the fraction of disk galaxies—from the empty field, to loose groups, to cluster outskirts, to cluster cores.

       3. Clearly, like humans, galaxy disks are sensitive to their environment.

   B. Simulations of rich galaxy clusters clarify how dangerous they are to delicate galaxy disks. There are frequent galaxy encounters that disrupt the disks.

   C. The gas in disks is also subject to "ram pressure stripping."

       1. Because galaxy clusters have a tenuous atmosphere, galaxies moving through the cluster experience a "wind" pressure. This can be strong enough to blow out the galaxy's interstellar gas, stripping it.

       2. There are a few examples in which the gas leaves a trail behind the galaxy.

       3. Generally, spirals in clusters have less gas than spirals in the field.

VI. What about our own galaxy, the Milky Way? What lies ahead for it?

   A. Our galaxy lives in a fairly empty region and hasn't experienced a major merger in 10 billion years.

   B. That will change in 3 to 4 billion years. Our neighbor, M31, is approaching at 70 miles/second and will begin a major merger that will last a billion or so years.

   C. There have been several attempts to simulate this merger, though the details of the incoming trajectory are still unknown.

D. John Dubinsky's simulation shows both an external view and a view from a star (the Sun?) participating in the collision. The night sky will offer some spectacular sights for our distant descendents!

Suggested Reading:

Hawley and Holcomb, *Foundations of Modern Cosmology*, chap. 15.

Silk, *The Big Bang*, chaps. 10–12, 15.

Jones and Lambourne, *An Introduction to Galaxies and Cosmology*, chap. 2.

Questions to Consider:

1. What happens when galaxies collide? Consider several situations: (a) when one galaxy passes by another, (b) when a small galaxy goes into orbit around a larger one, and (c) when 2 large spirals collide head-on.

2. Why does gas behave differently than stars during galaxy collisions? Why does some of the gas get hot while some gets cold, and what happens to each of these components? Why do galaxy collisions create so much star birth, and how is that visible to us observationally?

3. How are the disks of spiral galaxies made? Once made, what are some of the hazards facing spiral disks as the billions of years go by? Imagine following a spiral galaxy as it falls into a galaxy cluster. What effects do you expect to see, both on the gas in the disk and on the stars in the disk?

4. Imagine a million years pass like a second. Describe how our night-sky scene changes over the next 10 billion years (2½ hours—a movie's length of time). Include in your description the changing constellations, our orbit around the Milky Way, the approach of M31 and its eventual merger, and let's not forget the death of the Sun. Throughout all of this, does the Earth, as a planet, survive?

Lecture Twenty—Transcript
From Child to Maturity—Galaxy Evolution

In the last lecture, we left the story of galaxy development around 1 to 2 billion years after the Big Bang, equivalent to 5- to 10-year-old children. At that time, the galaxies were small and chaotic, a few thousand light-years across. Galaxy collisions and mergers were common. Star birth rates and supernova explosions were at an all-time high. It was a wild and spectacular time. In this lecture, I want to pick up where we left off and follow the story forward, 11 or 12 billion years to today. How did those little, blobby, dwarf galaxies mature into the large, majestic, modern-day galaxies, the beautiful spirals and ellipticals that we see all around us today?

The story is actually not unlike that for humans, in the sense that each of our lives has its own specific trajectory that takes us from child to adult, but there are also common events that most of us experience, like school, marriage, or a career. And it's like that for galaxies. Each has its own specific history, but there are common processes that many galaxies experience. So it's these common processes I want to look at in this lecture. And our very first process is a fabulous one—it's galaxy-galaxy collisions.

You should have been expecting this, because a fundamental part of the overall story is hierarchical assembly—small things merge to make larger things, which merge to make yet larger things. So let's look in more detail at what happens when galaxies collide. Well, just like a car accident, there is a range of possible collision damage, depending on whether the impact is a glancing blow or head-on, or on whether the collision involves 2 cars, 2 trucks, or a car and a truck. And sometimes, like on a foggy interstate, there can be a multicar—multigalaxy—pileup.

Now, the story of how we've come to understand what happens when galaxies collide spans 20 years and is a lovely example of a close collaboration between observations and computer simulations. So in this lecture, I'll be jumping back and forth between real images of galaxies and computer simulations. Now, the computational story got underway in the early 1970s, when two brothers, Alar and Juri Toomre, calculated what happens to the stars orbiting one galaxy when a smaller galaxy passes by, some distance away. Here's the output from their relatively simple computations.

The sequence unfolds with 50 million years between each step. As the small galaxy passes the larger one, its gravity affects the orbits of the stars, and they no longer travel on nice, neat circles but move out in larger arcs. And when you look at the whole collection, the pattern makes 2 huge, pretty, spiral arms, which astronomers call "tidal arms" because their formation is similar to the way the Moon's gravity raises tides on the Earth's oceans. Here are a few examples of real galaxies experiencing flybys. You can see very well this primary feature—2 arms, 1 towards and 1 away from the companion. It's a clear sign of gravitational tidal interaction.

Since those early days of limited computing speed and memory, things have improved enormously. Today's simulations routinely include 100 million stars, all pulling on each other, according to Newton's laws of gravity and motion. Actually, those 100 million computational objects are shared amongst the various galaxy components, not only the disk and bulge stars but also the dark matter particles that make up the heavy halo. Don't forget that a dark matter particle follows a trajectory just like a star would because it doesn't collide with anything. So it's very fortunate that the enormously important role of the dark matter can be included in the simulation as just another component.

Let's start gently with a fascinating process called "galactic cannibalism"—when a big galaxy slowly shreds and devours a smaller one. Here's a simulation of a small galaxy in orbit about a big one. The big galaxy's gravity gradually shreds the small one. The shredded stars enter similar, but not identical, orbits, which both lead and trail the small galaxy in what are called "tidal streams." After a few orbits, the streams extend right around the big galaxy, and in time the small galaxy may be shredded away to nothing. In effect, it's been eaten, and its stars have now joined the halo of the big galaxy.

Actually, what prompted these particular computer models was the discovery of half a dozen or so tidal streams around our own Milky Way galaxy. It seems we inhabit a galaxy with cannibalistic tendencies, although it's now thought that this steady ingestion of small galaxies by big ones is a routine part of a big galaxy's life. Here, for example, are 2 galaxies with prominent tidal streams in orbit about them.

Well, having looked at small galaxies interacting with large ones, let's now up the drama and look at a full-blown, head-on collision between 2 large spiral galaxies. So in this simulation, the disk stars are colored blue, the bulge stars are colored yellow, and the dark matter is colored red. The whole movie lasts just 15 seconds, but that covers 1.5 billion years, so time is passing at a rate of 100 million years per second—that's from the dinosaurs to now in just 1 second of movie time. So off we go. As they approach, there is some spiral structure induced in each disk, then at closest passage there is wholesale destruction of both disks, and fabulous tidal arms are flung out.

Now, I've paused the movie here to compare with a famous pair of interacting galaxies, called "the Mice." The agreement is remarkably good, though of course the initial setup conditions for this particular simulation were chosen to match this particular pair of interacting galaxies. Now, I hope you noticed how both galaxies effectively pass through each other, like 2 flocks of birds might fly right through each other. No stars ever collide. This makes sense when you remember our scale model from Lecture Two. If the United States is a galaxy, then stars are tiny human cells, separated by football fields. So when these galaxies collide, you can see that no cells, no stars will hit each other. All the effects that you see are caused by gravity gently modifying each star's trajectory.

Okay, let's continue the simulation. Notice how the relatively dense galaxy centers turn around and merge very quickly. You can really feel the strength of the central gravitational field. On the other hand, the outer parts evolve quite slowly, as those tidal features gradually wrap around and make, basically, quite a mess. The final state shows a single galaxy—one center, one bulge, and one halo. The 2 disks have been destroyed, and although some of the original disk stars have joined the final bulge, others make up a complex set of faint tidal features outside the main galaxy.

So now, let's look at some examples of real galaxies that resemble various stages of that calculation, from early to late. Of course, each is a different pair, but you can see in these examples many of the features that you saw in the movie. This last one certainly has all the hallmarks of a late-stage merger, complete with a complex set of outer tidal debris. Now, if you run these computer simulations longer, those outer halo features gradually get more and more washed out, like these galaxies here. And ultimately, the system resembles a nice smooth elliptical galaxy, like this one. This is a fascinating result. It seems that a galaxy's type—spiral or elliptical—can change as time goes by. At least some ellipticals once used to be spirals, until they got into a bad accident. So galaxies don't necessarily just change incrementally—occasionally, they undergo a sudden, total transformation.

Now, there's more to the story of galaxy collisions, because we've not yet considered what happens to the thin gas that exists between the stars, the interstellar gas. This is very important because stars are born from this gas, so we want to know what happens to it during the collisions. Let's begin with the computer simulations. Now, gas behaves in a much more complicated way than either stars or dark matter, and it wasn't until the 1990s that computers were sufficiently powerful to include it in the calculations. This sequence helps us see what happens. The left side shows the stars and dark matter. Actually, their motion is pretty much unaffected by the gas, and you can see the merger unfold from one panel to the next. However, unlike the stars, the gas from each galaxy doesn't simply pass through, but instead it collides at high speed, and the force of the collision both compresses and heats the gas.

What happens next is rather complex, but the upshot is that some of the gas remains hot and expands outwards. That hot gas is shown in the middle sequence, where the blue, green, and purple show the temperature from 20,000 K to 2 million K. However, some of the gas is able to cool, and it quickly gathers to the center of the galaxy. That's shown in the right sequence, where black, blue, and red show gas of higher and higher density. Now, I said before that gas can give birth to stars, so in all this chaos during the collision, are any stars born? The answer is an emphatic "yes." And it's not hard to see why. The compression resulting from the colliding gas clouds is enough to trigger the gravitational collapse to form stars. Also, the gas that goes to the center is now in a high-pressure environment that's quite conducive to forming stars.

So the upshot is that when galaxies collide, stars are born in huge numbers. Galaxy collisions are extremely fertile events. Now, observationally, the link between galaxy interactions and collisions and high star birth rates is very obvious indeed. For reasons I don't have time for, a classic symptom of high star birth rate is the emission of lots of far-infrared radiation. The first satellite telescope sensitive to this type of radiation was launched in the 1980s, and almost immediately, it became apparent that galaxies that shone brightly in the far-infrared, like these 4 here, were also in the throes of an interaction, or a merger.

These galaxies had huge amounts of gas in their nuclei and were birthing stars at a rate of 100 or even 1000 stars per year. That is much higher than the normal 1.0 or 0.1 new stars per year for a normal galaxy like the Milky Way. Galaxies with this kind of elevated star formation are called "starburst galaxies" because the episode of elevated star birth is usually quite short-lived. After about 100 million years, all the gas is used up or is blown out of the galaxy, so star birth shuts down. Perhaps the most famous example of a starburst is the pair of colliding galaxies called "the Antennae." And when the Hubble telescope zooms into the central regions, the image is shot through with the signatures of very high rates of star birth.

Now that we've begun to understand how gas behaves during galaxy collisions, we can also begin to understand the star birth history of the whole Universe. We met this topic last lecture, with the Madau plot. It showed the initial baby boom of high star formation, which is thought to be due to the high frequency of galaxy collisions in the young Universe. But could the later drop in star birth rate be due to a drop in the collision rate? Absolutely.

Using the Hubble telescope's deep surveys, you can measure the fraction of galaxies that are colliding, right across cosmic history. In the young Universe, almost all galaxies were colliding, while in the current Universe, only about 2% are colliding. This isn't surprising. As time passes, galaxies are getting further apart due to cosmic expansion, and in addition, they're combining, and so there are fewer remaining to collide. So it's pretty clear that at least part of the reason for the drop in cosmic star birth rate is due to the gradual decline in the number of galaxy collisions.

Let's now shift gears and look at how the primary parts of a modern-day galaxy gain their identity. The huge dark matter halo, a roughly spheroidal bulge of stars, and a relatively flat rotating disk made from both stars and gas—how were these components assembled? Well, in a sense, we've already looked at how the halos and bulges form. They're thought to be assembled during the initial collapse and ongoing collisions and mergers. If we use our merger tree from the last lecture, the halos and bulges are growing at these intersections during mergers. But what about galaxy disks? Disks are thought to grow during the relatively quiet times between mergers. They involve a more gentle acquisition of gas that falls in from the surroundings.

Here's a movie that shows the growth of a gas disk across several billion years. The movie shows only the gas, even though stars and dark matter are also in the calculation. Compared to the other movies, there's a different feel to this one. The collapsing objects rotate, and when they merge, the gas settles even more, although some is also heated and blown out. Even when modest collisions disturb the disk, it still seems to recover and continues to grow by pulling in gas. Finally, when the disk is relatively isolated, with no more mergers, it settles down, rotates, and even forms spiral arms.

Remember from Lecture Eighteen, the crucial difference between gas and either stars or dark matter is that gas is collisional and dissipative, meaning that it can't interpenetrate like stars or dark matter. It collides and compresses, and the energy of the collision can be radiated away as photons. It can be dissipated, and because of this, the gas settles down towards the center of the halo and forms a rotating thin disk. Now, although I've just described the formation of a rotating disk of gas, don't forget that over time, gas forms stars, which themselves inherit the orderly circular rotation of the gas. So over time, then, the disk builds up a population of stars.

That's the current situation with our own Milky Way galaxy. We've not had a major merger for several billion years, and during that time, both gas and stars have gradually built up to make a significant disk. In fact, 4.5 billion years ago, some of that gas condensed to form the Sun and planets. And here we are—the Sun has an approximately circular orbit, within the disk of the Milky Way galaxy. Actually, unmolested disks, left to themselves, have a gentle evolution of their own. As the hundreds of millions of stars orbit, they gently pull on each other, and although it's difficult to intuit what happens in this collective assembly, patterns form: spiral arms, rings, and bars. These may last a long time, or they may be transient, but they give their galaxies some of the most beautiful appearances in Nature.

Now, part of my story about disk construction was their need for relative isolation. Disks are fragile things and can be disturbed or destroyed relatively easily. Observationally, this is very obvious when you look at where spiral galaxies live, their environments. Here's the typical habitat of a spiral galaxy. Basically, they like to be alone, in relative isolation. Astronomers refer to this kind of location as the "field," meaning it's far from any major groups of galaxies. Now, look at a very different kind of galaxy habitat—a crowded cluster. Do you notice anything different about the type of galaxies inhabiting this cluster? There are essentially no spirals.

This fascinating tendency for spiral galaxies to shun dense environments was first put on a firm basis in 1980 by Alan Dressler. Here's a graph of his data, which nicely shows how the fraction of galaxies that are spirals gradually drops as you go from the empty field on the left, through loose groups to the periphery of clusters, to the centers of dense clusters. It's a steady decline. This graph really confirms that, like humans, galaxies are sensitive to their environment. It matters to a galaxy whether it lives out in the boonies or right down smack in downtown. So this now leads us to consider the environmental effect on galaxies.

First and foremost, spiral disks are fragile things and take time to build up. So in dense environments, spirals may simply not form, and if they do join a cluster later on, they may get destroyed by collisions. Here, for example, is a simulation of a high-density region. It starts about 2 billion years after the Big Bang and is destined to form a rich cluster of galaxies after about 8 billion years. Look how destructive the encounters are to the outer parts of the galaxies, tidally stripping away stars and flinging them out all over the place.

In these circumstances, a nice flat spiral disk has no hope of forming or surviving for very long. As the cluster steadily builds, it's really quite a chaotic place, with galaxies zooming around everywhere, getting shredded. Actually, if a spiral does fall into a cluster like this, one of the first things to be affected is its gas, the gas that's in its disk. It's removed in a remarkable process called "ram pressure stripping." When you stick your head out of a moving car's window, you feel a wind pressure. The air pushes on your face and blows your hair back. And if your hair were only weakly attached, it'd fly off. It would be ram pressure stripped.

Here's an example of that same effect, but on cosmic proportions. Here's a rich galaxy cluster, just like the one we saw modeled in the movie. But remember, these rich clusters also have a thin, hot atmosphere that shows up in X-rays, as you can see in this X-ray image of the same region. So if a spiral galaxy falls into a cluster like this, it feels a wind pressure blowing, and that's enough to strip the galaxy's gas away. You can see that's happening in this little galaxy here. It looks distorted, and it's leaving a trail of gas behind it as it zooms through the cluster at about 1000 miles per second. This process is, in fact, very widespread. Almost always, the spirals you find in clusters have less gas than spirals out in the field. It's a lovely example of environmental impact on galaxies. They really do feel the effect of their environment.

Let me end this lecture by looking at our own galaxy, the Milky Way. I mentioned earlier it seems we've not had a particularly dramatic life so far—no major mergers in almost 10 billion years. But that will all change in about 3 or 4 billion years' time. Here's our local environment, roughly to scale. Here's 1 million light-years. It's pretty empty. Here's the Milky Way, with its neighbors, M31 and M33, each with a handful of dwarf companions, and of course, let's not forget the dark matter halos that would look a lot bigger in comparison. Now, from Doppler shifts, we know the Milky Way and M31 are approaching one another at about 70 miles per second, and so it seems very likely that in about 3 or 4 billion years, they'll begin a major collision, which itself will last 1 billion or so years.

Not surprisingly, the drama and local nature of this event has led to a number of attempts to simulate the collision, though to be honest, since we don't yet know the sideways motion of M31, the details of the incoming orbit aren't really known yet. Probably the most detailed simulations have been done by John Dubinsky, so let's now watch his version of our distant future unfold. We start with M31 much closer, perhaps looking like this in the sky. Here's a fish-eye view of the whole sky. Around the rim is the horizon, and the Milky Way disk crosses the sky. John's movie adopts this view. Here it is, with time flying by at millions of years per second. You can see some of the hundred million stars in the calculation zipping across the night sky.

Let's now jump to an external viewpoint to get the big picture. The 2 galaxies are about to collide. How we witness the interaction depends, of course, on which star you pick to be the Sun. In this case, John picked a star that gets flung way out, so that in time we'll look back as the interaction proceeds. Also, be on the lookout for us falling back in and passing right through the galaxy center. Remember, no stars ever collide, because they're so tiny compared to their separations—so the Sun and Earth will probably survive this whole encounter.

So with this gentle, billion-year drama continuing to unfold, let me summarize what we've learned in this lecture. My aim was to look at ways in which galaxies have matured over the last 10 billion years. As always, interactions are a prime player, so we looked in detail at some examples of these. Galaxy flybys create S-shaped tidal arms, while dwarf galaxies falling into and orbiting a big galaxy can get slowly shredded and eaten as their stars get pulled into long tidal streams. When 2 spirals collide head on, spectacular tidal tails and bridges are formed first. The central regions quickly merge, while the outer parts evolve more slowly, initially as a big mess, and ultimately as a smooth elliptical galaxy.

The gas within colliding galaxies undergoes a rather different response. Some is heated and leaves the galaxy, but much is compressed and cools and falls to the galaxy center, where it experiences vigorous star birth. Simulations suggest that present-day halos and bulges each form from collisions and mergers, but disks seem to grow in quieter times, as gas falls in and settles down. Over time, this gas gets converted into stars, and gravitational interactions within the disk can give rise to beautiful patterns, with spirals, bars, and rings. Disks are, however, rather fragile things and don't survive long in dense environments, which is why spiral galaxies are rarely found within rich galaxy clusters. When they do fall into a cluster, the first thing to go is the gas, stripped out by ram pressure against the cluster's thin, hot atmosphere.

All these processes, and several more I've not had time to mention, play out their role in shaping the present population of galaxies, changing them from the small, amorphous infants we met in the last lecture into the huge, stately masterpieces we see all around us today. But our story of galaxy maturation isn't over. There's another, darker side yet to come. It involves the growth of massive black holes in galaxy centers. It involves their devouring of stars and gas, and it involves their shutdown of galaxy construction, but that's the next lecture.

Lecture Twenty-One
Giant Black Holes—Construction and Carnage

Scope:

Work in the 1990s showed that essentially every galaxy, including our own, has a massive black hole sitting at its center. These are fascinating astrophysical beasts, with Herculean gravitational fields. While most are invisible, others capture gas from their surroundings and light up like beacons, making the galaxy "active." Because galaxy collisions send gas to galaxy centers, active galaxies were much more common in the Universe's youth, when galaxy collisions were frequent. Black holes also play an important role in the story of galaxy growth. When we try to understand the whole spectrum of galaxy masses, we find four processes that suppress the formation of really big and really small galaxies, and black hole energy output is one of these four. We'll end by looking at the latest cosmological simulations that include all these effects. They reproduce, surprisingly well, many of the important properties of the entire galaxy population.

Outline

I. In this lecture, I want to introduce a very different kind of character to our story: massive black holes. These powerful objects are fascinating in their own right and play a significant role in the lives of galaxies.

  A. This subject got a huge boost in 1997 when a new spectrograph was installed on the Hubble Space Telescope, allowing astronomers to measure Doppler shifts of stars and gas very close to galaxy centers.

  B. Very high orbit speed that is very close to the center indicates the presence of a large mass, much more than the visible stars or gas.

  C. The orbits of individual stars in the Milky Way's center reveal the presence of a black hole weighing 3.7 million times the Sun's mass.

D. Two key facts have emerged.

    1. Essentially all galaxies have massive black holes at their center. The only exceptions are galaxies with no nucleus or bulge.

    2. Galaxies with bigger bulges have more massive black holes.

II. The primary quality of a black hole is that a great deal of mass fits in a tiny space, so that the gravity nearby is extremely strong.

    A. If the Earth were a black hole, it would be ½ inch across!

    B. The gravitational energy released when a mass, m, falls into a black hole is roughly $E = mc^2$, so black holes can be incredible sources of energy.

III. Let's consider black holes in galaxy nuclei.

    A. A black hole weighing 50 million solar masses is as big as the Earth's orbit. This is tiny relative to the galaxy: a pinhead in the middle of the United States!

    B. Besides orbiting around black holes, stars are unaffected by them. Such black holes are invisible.

    C. However, if gas falls into the galaxy center, it can settle into an "accretion disk," which itself feeds the black hole.

        1. Gas moving through the accretion disk releases huge amounts of energy, and the nucleus lights up like a beacon.

        2. We call this an "active galaxy."

    D. The energy released emerges in three different ways.

        1. The hot disk glows brightly: from infrared to X-ray.

        2. A wind blows off the disk.

        3. Two jets of thin gas are launched at high speed perpendicular to the disk.

    E. Depending on the amount of energy released and how it shows up, there are various kinds of active galaxies: Seyfert galaxies, quasars, and radio galaxies.

    F. Sending gas to the center—"feeding" the black hole—occurs when galaxies collide. Simulations of colliding galaxies nicely show this.

IV. Let's view galaxy black holes with a cosmological perspective.

    A. If galaxy collisions trigger black hole feeding, we should find active galaxies to be more common in the past, when galaxy collisions were more frequent.

    B. The observed quasar history shows exactly this: a rapid turn-on after a billion years, a heyday at 2–3 billion years, and then a steady decline. At their peak, quasars were 10,000 times more common than today.

    C. In today's calmer Universe, black holes lay dormant, unfed. Occasionally, galaxy interactions send gas down for a snack, and mild activity resumes.

    D. The feeding history of black holes is also their growth history. How do you build a 1-billion-solar-mass black hole?

        1. Feed a seed black hole at 1 Sun mass per year for a billion years.

        2. That's about 1 Earth mass per 2 minutes. Using $E = mc^2$ and 10% efficiency, we get 10^{39} watts of power, or 100 times brighter than a galaxy—a typical quasar power.

    E. How are the initial seed black holes created? It's not yet known, but it probably happens during the birth of the first stars or first galaxies, because quasars were shining after just a billion years.

V. Let's now see how black holes affect galaxy growth, which doesn't just occur via hierarchical merging. Another process is called "feedback"—black hole energy output can shut down galaxy construction.

    A. Let's start by considering the "mass distribution of galaxies"—the relative numbers of low-, medium-, and high-mass galaxies.

    B. Perhaps this matches the mass distribution of dark matter halos, because stars are thought to form at the centers of halos.

    C. Giant computer simulations of dark matter reveal that there are very few big halos, more medium-sized halos, and lots of small halos.

    D. The data for real galaxies is both similar and different.

        1. The overall distribution is similar: few big galaxies, many little ones.

      2. But the match is bad for very massive and very low-mass galaxies. There are way fewer galaxies than dark matter halos.

      3. Stated differently, we don't see galaxy clusters with one huge galaxy, and we don't see normal galaxies with lots of dwarf companions.

  E. Why the difference? Because making a galaxy requires *two* things.

      1. A dark matter halo must form.

      2. Atomic gas must settle into the halo and form stars.

  F. Whether the second step can actually occur depends on the strength of the halo's gravity, and that depends on its mass.

      1. For huge halos, the energy released by falling in makes the gas so hot it can't form stars. Rather than making a huge galaxy of stars, we have instead a huge, hot atmosphere.

      2. For big halos, the energy injected by black holes in active galaxies heats the infalling gas, which again inhibits stars from forming.

      3. For small halos, with weaker gravity, the first few stars and supernovae blow the gas out, leaving a halo with no gas or galaxy.

      4. For tiny halos, the weak gravity can't overcome the pressure of the surrounding gas, which never falls in.

VI. Let's now fold these mechanisms into our simulations of galaxy formation.

  A. Because they involve complex gas physics, they are inserted into the dark matter simulations using simple formulas to make what is called a "semi-analytic model."

  B. These hybrid simulations now reproduce many observed properties of the galaxy population.

      1. They get the history of star formation correct: an initial baby boom, followed by steady decline.

      2. They find more elliptical galaxies in dense environments.

      3. They match the galaxy mass distribution extremely well.

      4. They get the quasar history correct: many early; few today.

      5. They find bigger black holes in more massive galaxies.

C. Although this is still a young subject, it seems we're well on the way to understanding the birth and development of the entire galaxy population.

Suggested Reading:

Hawley and Holcomb, *Foundations of Modern Cosmology*, chap. 9.

Silk, *The Infinite Cosmos*, chap. 7.

Jones and Lambourne, *An Introduction to Galaxies and Cosmology*, chap. 3.

Begelman and Rees, *Gravity's Fatal Attraction*.

Questions to Consider:

1. How are massive black holes found in galaxy centers? What changed in the mid-1990s that made their discovery so much easier? Why is the evidence for a black hole at the center of the Milky Way so convincing?

2. Why can black holes be such powerful sources of energy? Why do most black holes in galaxy centers go unnoticed, while a few are immensely bright? Why were luminous black holes more common in the past?

3. Compare the shapes of the following three mass distributions: humans, galaxies, and dark matter halos. How do we know the shape for halos? Why are the shapes for galaxies and halos different? What suppresses the formation of extremely massive galaxies inside cluster-sized halos? Conversely, what suppresses the formation of dwarf galaxies within small halos?

4. Imagine if the Sun (mass $= 2 \times 10^{30}$ kg) were torn apart by a black hole and, via an accretion disk, devoured in a year (a quite plausible fate for a star in a galaxy center). How brightly would this final "blaze of glory" shine? Hint: Use $E = mc^2$ and 10% efficiency to derive the total energy release (in joules, use $c = 3 \times 10^8$ m/s) and divide this by the number of seconds in a year to convert to watts. [Answer: You should get about 5×10^{38} watts, or 10^{12} L_{Sun}—100 times brighter than a typical galaxy!]

Giant Black Holes—Construction and Carnage

So far, our story of the growth and development of the Universe has included relatively familiar things—stars, galaxies, gas. And even though dark matter isn't exactly familiar, perhaps you're beginning to get a feel for it. But this lecture, I want to introduce a very different kind of character. If this was a traditional creation story, I'd be introducing a dragon or a demon or some ungodly destructive power. But in the real Universe, this ungodly power takes the form of massive black holes. At the heart of essentially every galaxy lies between a million and a billion Suns of matter, crushed into a region smaller than the solar system, whose powerful gravitational field causes essentially everything within its vicinity to zoom around at almost light speed. It really does evoke the idea of a fire-breathing dragon, hiding in its lair.

So in this lecture, I want to tell you some of the story of these massive black holes and how they play a role in affecting the lives of galaxies. Let's begin with the amazing observational progress in finding massive black holes at the centers of galaxies. The subject took a huge step forward in 1997, when the Hubble Space Telescope had a spectrograph installed that allowed astronomers to measure, using Doppler shifts, the movement of stars and gas very close to galaxy nuclei. Here's an example. Hubble zeroes in on the center of this elliptical galaxy in the Virgo Cluster, and then places the tiny spectrograph aperture across the nucleus. At each point along the aperture, a spectrum is made, which shows the Doppler shift of the orbiting gas and stars.

In this case, within just 25 light-years of the nucleus, the gas is orbiting at over 400 kilometers per second. You can see also the drop in rotation speed away from the nucleus, just like for the solar system, showing that the gravity comes from a single mass at the center. Applying Newton's laws, just as we did in Lecture Nine, you can measure a mass of 300 million solar masses, far more than is present in the actual stars. Now, since then, similar observations have been made for nearly 100 galaxies, and in essentially every case the Doppler shifts indicate a massive object right at the center.

Even our own Milky Way galaxy has a central black hole. Here's a beautiful nighttime view of the Milky Way, with its center located right here in the constellation of Sagittarius, smack behind a lot of obscuring dust. Now, fortunately, if we look using infrared light, like this image, you can see through the dust. Zooming in, you see a central star cluster, and 2 groups, 1 in Germany and 1 in the United States, have carefully followed the motion of the stars in this cluster.

Here's a movie in which 15 years pass in 5 seconds, and you can see nice, neat, elliptical orbits around a central point. Here is it, repeating. Try to watch star 14 carefully. At closest approach, star 14 is moving at 3000 miles per second and comes within 6 light-hours of the central point. That's as close as the planet Uranus is from the Sun. Careful analysis of the orbits essentially proves that sitting at the central point is a black hole, with 3.7 million times more mass than the Sun.

Now, as I said, for almost 100 galaxies, black holes have now been found, and their masses measured. And there are two remarkable facts that have emerged. First, essentially, every galaxy that's been looked at has revealed the presence of a black hole. The only exceptions are galaxies with no real nucleus or bulge, things like our local dwarf galaxies, the Large and Small Magellanic Clouds. Otherwise, massive nuclear black holes are a normal galaxy component, a part of its anatomy, like an arm or a leg. And secondly, as this graph shows, the black hole mass up the y-axis increases with the galaxy bulge mass. That's along the x-axis. For example, the huge galaxy M87 has a 3-billion-solar-mass black hole, while shown at the same scale, the lightweight galaxy M32 has a lightweight 200-million-solar-mass black hole. And on average, the black hole mass is about 1/1000 of the mass of the galaxy bulge.

Now, at this point, I should say something about the nature of black holes. From our perspective in this course, their main quality is simply that they contain a great deal of mass within a tiny size, so that close to the black hole, the gravity is extremely strong. Just to give you an idea, here's the Earth, with about 10^{25} kilograms of rock within a sphere 8000 miles across. So near its surface, gravity pulls with modest force. We have little arrows. If I drop a bag of flour through 1 foot, the energy it releases sounds like a thud—not very much gravitational energy released there. But I crush the Earth down to make a black hole—it's now half an inch across—with the same mass, 10^{25} kilograms. So the gravitational force nearby is huge. We have huge arrows. Now drop a bag of flour from 1 foot—it falls in about a nanosecond, moving at almost the speed of light, and the thud of its landing is equivalent to detonating a large hydrogen bomb.

In fact, the gravitational energy that gets released when matter falls into a black hole is roughly given by Einstein's famous $E = mc^2$, so black holes are potentially incredible sources of energy. Feed them anything, and you can convert the infalling mass almost directly into energy. Now, in practice, the efficiency isn't quite 100%. It's more like 10%, but even so, that's enormously more efficient than, for example, nuclear energy at 0.1% or chemical energy at one-millionth of a percent efficiency.

Now, back to the astrophysical context: a galaxy nucleus with a black hole, weighing, let's say, 50 million solar masses. At that mass, its size is about the size of the Earth's orbit. This is tiny relative to the galaxy. From our scale model in Lecture Two, it's a pinhead in the middle of the United States—around this pinhead, stars the size of human cells 10 yards away, where the black hole's gravity is much weaker. So for most of the time, the black hole just sits there and is, well, black. You don't see it. This is the condition of the black holes I just showed you in that elliptical galaxy and in the Milky Way. It's all but invisible, although its gravity affects the orbits of stars out to a few light-years, and that of course is how we know it's there.

There is another situation that can occur. If gas can fall into the very center, close to the black hole, then it settles into what is called "an accretion disk." It gradually spirals inwards, orbiting faster and faster as it gets closer and closer to the black hole, until it's basically moving at light speed just before it falls in. As the gas moves through this disk, before it finally enters the hole, it can release huge amounts of energy, and the galaxy nucleus lights up like a beacon. Astronomers call this an "active galaxy." It's a galaxy whose central black hole is having a meal—it's devouring gas and stars, and releasing energy as it does so.

Now, exactly how this accretion disk behaves isn't, in fact, understood very well, but we think that there are three primary ways that energy is released:

- The hot disk is extremely bright. It produces photons, from infrared to X-rays.
- A wind blows off the disk.

- Two fast-moving jets of very thin, but very hot, gas are launched perpendicular to the disk.

Let's have a quick look at some active galaxies. They go by various names—quasars, Seyfert galaxies, radio galaxies—depending on just how the energy makes itself known and how much energy there is. Here, for example, are some quasars. Their active nucleus is so bright that one can barely see the host galaxy. This tiny region, smaller than the solar system, is producing more light than all 100 billion stars in the galaxy combined.

Here are some active galaxies whose black holes are driving out powerful jets. These jets produce radio emission, which the images show as red, while the visible light from the galaxy of stars is in blue. So you can see how the jets push way out, well beyond the galaxy, out into the galaxy's halo. Not surprisingly, these are called "radio galaxies." Now, you may have noticed something interesting, particularly about the quasar images. They seem to be disturbed or interacting galaxies, and this gives us a clue about what triggers activity, what initiates feeding the central black hole. Remember from the last lecture, when galaxies interact, some of their gas falls to their center.

Here's a movie of 2 galaxies colliding, to remind us—starting with just the stars and dark matter. Now, here's the same collision, but showing just the gas component. Notice how it ends up in a much more compact region than the stars. And here, let's watch that last part again, zoomed in a bit. The 2 gas disks merge, and as they do so, the gas really settles to the center. So these simulations show how "lunch" gets transported down to feed the central black hole, and so the galaxy becomes active. Now, let's begin to put all of this into a cosmological perspective.

In previous lectures, we've seen how galaxy interactions were much more common in the past, and that, in turn, led to much higher star birth rates in the past. Well, since feeding black holes also seems to be triggered by galaxy interactions, do we find evidence that activity in galaxies was also greater in the past? And the answer is a very definite "yes." Here's the distribution of 10,000 quasars, very powerful, active galaxies, in one of those pie diagrams, where we are at the center. Look at the distribution. There are virtually no quasars nearby, whilst there are thousands at greater distances. Now, look at the time axis. We go right out to 14 billion light-years, essentially all of cosmic history. Clearly, quasars were much more common in the past compared to today.

Here's the final analysis that shows the rise and fall of black hole feeding. There is a rapid turn-on after about a billion years, and a heyday between 2 and 3 billion years, when powerful quasars were 10,000 times more common than they are today. Since then, quasar activity has gradually shut down as galaxies have settled into a more sedate and tranquil middle age, starving their central black holes. But just occasionally, in today's Universe, galaxy interactions send a bit of gas down, a snack for the dormant black hole, and the activity resumes once more, but at a more modest level. We call these lower-power active galaxies "Seyfert galaxies" after the American astronomer, Carl Seyfert, who first studied them in the 1940s.

Now, unlike humans, when black holes eat, they don't excrete. This is a one-way trip for the food, so all this discussion of feeding is equivalently a discussion of their growth, of their acquisition of mass. After all, how do you think a black hole weighing 1 billion solar masses actually got there? Well, maybe you feed an initially small black hole at a rate of 1 solar mass per year for 1 billion years. Let's just check to make sure this is reasonable. One solar mass per year is equivalent to about 1 Earth mass every 2 minutes. Now, if you convert that mass to energy, $E = mc^2$, at 10% efficiency, you get 10^{39} watts of power. That's about 1000 billion times the power of the Sun, or 100 times brighter than the Milky Way galaxy. Yes, that's about how bright quasars shine, but that's just their growth. How do you get those initial seed black holes, which then feed and grow?

Well, at the present time, we don't know for sure. Maybe they formed when the first stars formed, or maybe later, when the first galaxies were assembled. But either way, it must have been fairly rapid, because the most distant quasars were shining brightly just 1 billion years after the Big Bang. Now, at this point, I want to broaden our discussion of black holes. So far, I've talked about them because they're a part of almost every galaxy, and they're interesting in their own right. But let's not lose sight of our overall story here. We're trying to explain how the population of galaxies came to be the way it is, and so far, we've really only considered one main plotline: Galaxies develop by gravitational interaction and hierarchical merging.

But in the last 10 years, we've begun to recognize that this cannot be the whole story. There is a second plotline, and it goes by the rather prosaic term "feedback." Basically, the huge energy release from galaxy centers, due either to black holes or high star birth rates, can severely affect the galaxy and its environment, and in some circumstances it can shut down star formation and inhibit the further construction and growth of galaxies. Returning to our dragon metaphor for a moment, despite being confined to its lair, the dragon's fiery breath scorches the surrounding land, preventing anything from growing.

So let's now locate this new theme of feedback in its broader context of galaxy growth. And to do that, we need to consider what's called the "mass distribution of galaxies." Now, here, for example, is a sketch of the mass distribution of humans. It has a peak around 200 pounds, a short tail above for obese adults, and a longer tail below for children and infants. It gives a pretty good feel for the human body. We're way bigger than mice or ants, and we're smaller than elephants or whales. So what do you think is the equivalent graph for galaxies, where I'm meaning the stars now, not the dark matter? Is there a maximum or a minimum galaxy mass, and if so, why?

Now, I'm going to keep you in suspense because I want to first show you a very plausible guess. Remember that hierarchical merging builds a whole population of dark matter halos into which atomic matter falls and makes stars. So maybe each dark matter halo contains a galaxy, so we would expect a galaxy mass distribution to be the same as the halo mass distribution, but perhaps shifted a bit because there's about 6 times less atomic matter than dark matter. So with this idea, we need to know the mass distribution of dark matter halos. Now, currently, we can't measure that observationally, but we think computer simulations are sufficiently trustworthy to provide the answer. I'll be discussing these huge cosmological simulations in the next lecture, but for now, let's just go straight to the results.

Here's what they calculate for a typical distribution of dark matter across a huge region, spanning hundreds of millions of light-years, containing hundreds of thousands of dark matter halos. As you can see, there are relatively few big halos, more medium-sized halos, and lots of little halos. When you count them all up, you get this graph— the mass distribution of dark matter halos. Notice the axes are exponential and go from huge, galaxy cluster–sized halos of a million billion solar masses right down to tiny dwarf galaxy halos of just a billion solar masses. As is so often the case in astronomy, there are way more small things than big things, and above some threshold, nothing bigger is found.

Now, allowing for atomic matter being 6 times lighter than dark matter, we might expect a galaxy mass distribution like this. So what's the real observed galaxy mass distribution? Are you ready? Here it is. Well, the good news, I suppose, is that they do match for Milky Way–sized galaxies, and the overall shape is similar—many more small galaxies than big ones, with a steep drop in the number of big galaxies. But the bad news is that the really big halos simply don't contain really big galaxies, and there are far fewer dwarf galaxies than expected.

So we're beginning to get the picture here. To make galaxies, two things must happen: First, you must make the dark matter halos. That's okay—we understand that. But then you need the atomic gas to settle into those halos and form stars, and that step isn't straightforward. It's vulnerable to several processes that can inhibit it. Now, the reason galaxy formation is suppressed has all to do with the depth of the gravitational field of the dark matter halo, how strongly it pulls on the atomic matter. Here's a sketch, illustrating the pull by the size of a valley into which the gas is falling.

For the biggest halos, the energy of falling in heats the gas so much that it can't cool to form stars. These enormous-sized halos make clusters of galaxies. Remember from Lecture Nine, though, these also have huge, hot, X-ray-emitting atmospheres. There's plenty of atomic gas—it's just way too hot to form a huge galaxy. Of course, preformed galaxies do fall into the halo, and we have the familiar galaxy cluster—the Coma Cluster in this example. Now, in slightly smaller halos, where the gas could cool, energy coming out from big black holes in active galaxies is thought to keep the gas hot, so once again, it can't cool to form stars.

Here's an example of a galaxy with an X-ray atmosphere shown in blue and a black hole driven jet shown in red. Notice how the jet has pushed 2 big holes in the atmosphere, which heats it and prevents it from cooling. In even smaller halos, the problem is holding onto the gas. Gravity is so weak that when the first stars and supernovae form, they blow the gas back out, and this halts the formation of a small galaxy. Actually, we see this sort of thing happening nearby, with starburst galaxies.

Here's a lovely image of the nearby galaxy M82. It has loads of star formation and supernovae, and you can see how they're blowing out the gas in these great plumes above and below the galaxy. Now, finally, for the smallest halos, the gravity is so weak it can't even pull in the gas, because the gas has pressure, and this prevents it falling in—just as the air in this room doesn't fall towards the weak gravity of my body. Let's now see what happens when we fold all of these mechanisms into our understanding of galaxy formation.

As you might imagine, these four mechanisms are complex, messy things. We're barely able to model them for a single galaxy, let alone a million galaxies in a giant cosmological simulation. So to make progress, these processes are inserted into the giant simulations using a shortcut method, and the resulting hybrid is called a "semi-analytic model." Basically, a giant simulation is run for the dark matter, but the complex gas processes are added, using simple formulae that describe things like the fraction of gas-forming stars or the efficiency with which supernovae blow the gas out, and so on. In all, there are about 10 such formulae.

So let's look now at the population of galaxies that emerge from these semi-analytic models. Here you can see, in gray, the dark matter distribution that provides the framework within which galaxies might form. Now, the colored dots are where the galaxies are forming. Here, at just 2 billion years after the Big Bang, they're mainly blue, which signifies they have high star birth rates, so many young stars are present. Let's come forward in time. More galaxies form, and in many, the gas is blown out, so star birth ceases—so the population of stars age. And that aging population is color-coded— green, orange, and red for the oldest population. By today, we have a population of galaxies with lower overall star birth rates. They're more yellow and red, and the red galaxies—these would be the ellipticals—are found in the densest regions. So the model certainly seems to match two key results from previous lectures: the decrease in overall star birth rate and the tendency of galaxies in dense regions to be ellipticals.

Now, what about the galaxy mass distribution? Well, here's our previous graph of the halo mass distribution in red and the observed galaxy mass distribution in blue, and here are the results of the semi-analytic model in green. Now, that's an impressive match, although it isn't quite as impressive as you might think because, don't forget, some of the model's free parameters were chosen to optimize the match. But even so, I was impressed when I first saw this. There are other things these models get correct. They get the history of star formation and quasar activity roughly right: early wild times, followed by a steady decline. They also get the correlation between bulge mass and black hole mass correct because the same conditions that build bulges also feed black holes and cause them to grow.

Well, I think at this point you've got the picture. We're beginning to explain in broad terms quite a range of key galaxy properties. It's still early days in this subject area, and in the coming years you can expect to see significant developments, both in the observational story and its parallel, the theoretical and computational story. We've covered a lot in this lecture, so let me quickly summarize the main points.

Since the installation of the spectrograph on Hubble, we now realize that essentially all galaxies with a bulge also have a massive black hole. In today's Universe, most of these black holes are invisible, except for their influence on the motion of stars and gas nearby in the nucleus. However, if gas falls into the black hole, the enormous drop liberates approximately 1/10 of the mass of infalling material as energy, and we say the galaxy is "active." Of course, galaxy interactions cause this feeding of gas to the center, so not surprisingly, we find activity to be more common in the Universe's past, when interactions were common.

The enormous energy coming from galaxy nuclei can significantly influence subsequent galaxy formation, a process known generally as "feedback." One way to uncover the effects of feedback is to compare the mass distribution of dark matter halos with the measured mass distribution of real galaxies. And when you do this, it's clear that something is preventing the really huge and really small galaxies from forming. We think truly giant galaxies can't form—in part because the gas in huge halos is too hot and thin to cool and settle, and in part because jets driven by black holes keep the gas hot and prevent it from cooling. And we think tiny dwarf galaxies can't form, in part because supernova explosions drive the gas out of the galaxy before it can properly form.

When all these effects are included in the giant cosmological simulations, many of the observed properties of galaxies are found, and although it's still early days in this particular subject area, it seems we're well on the way to understanding, at least in broad outline, how the population of galaxies has come to be the way it is. But these are all still only the internal properties of galaxies. We have one final and critical aspect to look at, and that is the location of the galaxies—how they fill the Universe, the patterns that they form. And that's the topic for our next lecture.

Lecture Twenty-Two
The Galaxy Web—A Relic of Primordial Sound

Scope:

How are galaxies distributed throughout space? Red shifts of a million galaxies provide maps that show great voids, walls, filaments, and clusters—and even reveal a faint imprint from primordial sound waves. Gigantic computer simulations convincingly show how these galaxy patterns come about as gravity acts on the roughness emerging from the Big Bang. The simulations also show how galaxies move from voids to filaments, then down these to build clusters. It's even possible to see these galaxy flow patterns in the real Universe by carefully combining distance and redshift measurements. We look at a map of our own galactic neighborhood—out to 300 million light-years—which includes several superclusters. We're "falling" toward the Virgo Cluster at 300 km/s, which is itself falling toward the Great Attractor at 500 km/s. We will, however, never arrive at the Virgo Cluster: Dark energy's accelerating expansion will soon halt any further structure growth.

Outline

I. This is the last of our six lectures on the growth of cosmic structures. We've spent most of our time looking at the growth of galaxies, but now I want to step back and look at the pattern of galaxies across space.

  A. If we want to know the 3-dimensional pattern of galaxies, we need to measure their distances. With so many, this is too difficult, and instead red shifts are used to infer distance assuming the Hubble law.

   1. Measuring galaxy red shifts was pioneered in the 1970s and '80s by a group at the Harvard Smithsonian Center for Astrophysics. They measured 18,000 red shifts one by one.

   2. Today, spectrographs take hundreds of spectra at one time.

   3. Two huge projects have recently measured a million red shifts.

B. One way to show the patterns is to take a strip of sky with about 100,000 galaxies and plot it as a pie-slice, with us at the apex and a dot for each galaxy.

　　1. These maps reach 2 billion light-years, 1/7 of the way to the visible limit, so this is a significant fraction of all there is to see.

　　2. The galaxy pattern has a mottled, weblike appearance, with three types of features: voids, filaments that border the voids, and clusters where the filaments intersect.

C. This kind of pattern emerges when gravity acts on the primordial roughness that emerged from the Big Bang. The best way to show this is to use computer simulations that recreate the growth of large-scale structure.

II. Let's follow one particular simulation from 2005. It tracked the gravitational motion of 10 billion dark matter "particles" spread throughout a 2 billion light-year cube. After 14 billion years (28 days of computer time), they'd formed 18 million dark matter halos, of which 2 million contained galaxies.

A. A movie takes us through the final computational volume. There is a clear sense of the 3-dimensional web made from dark matter halos. We also visit a huge cluster of galaxies at the intersection of several filaments.

B. Comparing with observed galaxy maps shows remarkable similarity. Clearly, our understanding of structure formation is on the right track.

C. The simulation also allows us to watch the web evolve over time.

　　1. The initial distribution (at 200 Myr) is fairly smooth, as you'd expect.

　　2. Quite rapidly, matter moves away from the lower-density regions to the higher-density regions, enhancing the density contrast.

　　3. The web of filaments emerges quite quickly and continues to grow more prominent.

　　4. Galaxies appear to flow down filaments onto clusters. For them, "downhill" is toward the cluster.

D. Filaments form between clusters because there is a linelike enhancement of density, and it collapses inward to make—in 3 dimensions—a filament.

III. Let's briefly return to our description of structure using the roughness spectrum from Lecture Seventeen. Recall, it's the spectrum of waves that, when summed, make the density pattern.

    A. The Universe's roughness spectrum can be measured using the CMB and the galaxy maps. There is a peak around 200 Mly, and it falls to either side—you can find ever-larger structures up to 200 Mly, but beyond that the Universe gets ever smoother.

    B. There is, however, an additional feature in the spectrum: slight wiggles on the short-wave side of the peak. What are they?

    C. They're a relic of the acoustic harmonics from before 400,000 years. The sound wave patterns visible on the CMB survived through galaxy construction and are now visible in the roughness spectrum, where they're called "baryon acoustic oscillations."

    D. This is wonderful: Primordial sound is still with us in the patterns of galaxies.

IV. It is also possible to measure the flow patterns of galaxies as they respond to the gravity of their neighbors.

    A. Remember, the galaxy maps are made using red shift. For a perfect Hubble law, that should convert to a perfect distance.

    B. But galaxies also move in response to their neighbors, and this velocity gets added to their measured red shift.

    C. If we measure their *true* distance (using, e.g., the Tully-Fisher method), then we can figure out this additional velocity.

    D. Overall, combining accurate distance measurements with redshift maps can give you a galaxy flow pattern.

V. Using this approach, let's look at our local neighborhood and its galaxy flow pattern. We'll take three views: medium, small, and then large.

    A. First the medium perspective: out to 300 Mly.

        1. Several superclusters are visible, including Perseus-Pisces, Centaurus, and the Great Attractor.

        2. The measured flow seems to carry galaxies away from voids and toward denser regions, just as we expect.

   3. The Milky Way itself is falling toward the Great Attractor at roughly 500 km/s (relative to the Hubble expansion flow).

B. Next a more local perspective: out to 100 Mly.
   1. The Local Group sits in a small filament that connects to the Virgo Cluster, which has 2 larger filaments entering from either side.
   2. The Local Group is falling toward Virgo at 300 km/s, while this entire region is falling toward the Great Attractor at 500 km/s (relative to the Hubble expansion flow).
   3. Again, the overall flow is away from voids and toward denser regions.

C. Finally, let's step way back to include the surrounding 800 Mly.
   1. In addition to the previous clusters, we now include the Coma Cluster and the huge Shapley concentration 500 Mly away.
   2. These are, however, too distant to affect us. Our motion comes mainly from the pulls of the Virgo Cluster and the Great Attractor.

VI. Finally, let's look at the future of structure formation.

A. In the picture of hierarchical assembly, small regions turn around first, then larger ones. The measured flow patterns are simply the latest stage: Superclusters are beginning to slow their expansion. Will they ever turn around and collapse?

B. No, because dark energy is now driving an accelerating cosmic expansion, and this halts any further growth of structure.

C. A simulation of the distant future shows little change from today.
   1. We will never join the Virgo Cluster.
   2. An event horizon forms where expansion reaches light speed. Galaxies outside this become invisible.
   3. Ultimately, even Virgo passes outside the event horizon, and our (then merged) galaxy is fundamentally alone.

Suggested Reading:

Hawley and Holcomb, *Foundations of Modern Cosmology*, chap. 15.

Ferris, *The Whole Shebang*, chap. 6.

Silk, *The Infinite Cosmos*, chap. 11.

Dressler, *Voyage to the Great Attractor*.

Rees, *Just Six Numbers*, chap. 6.

Bartusiak, *Archives of the Universe*, chap. 72 (CfA paper).

Questions to Consider:

1. How do we measure the 3-dimensional distribution of galaxies over a sufficiently large volume to see their patterns? What are those patterns—how are galaxies spread throughout space? Why are there voids, clusters, and filaments?

2. Even though the galaxies appear frozen on the sky, we can in fact figure out how they "flow" as they're pulled by the surrounding matter. How is this possible? Ignoring Hubble expansion, what is the overall flow pattern of galaxies, and does it match what you expect for gravitational motion?

3. With nothing to hold the Milky Way in place, it is "free" to fall wherever gravity tells it to. In what direction is the Milky Way currently falling? Consider forces within the Local Group, within 50 million light-years (which includes the Virgo Cluster), and within 200 million light-years (which includes the Great Attractor).

4. Compare the ability of galaxies to clump up and form structures in three different situations: (a) a static, nonexpanding Universe (rather like Newton envisioned); (b) an expanding but decelerating Universe like we find during the matter era; and (c) an expanding but accelerating Universe, like we find in the dark energy era. What implications does our transition into the dark energy era have for the future development of cosmic structure? Will we ever join the Virgo Cluster?

5. Imagine a future race of sentient cosmologists, living 200 billion years from now, orbiting a star near the then-dead Sun. Describe what their telescopes would reveal to them. What would the "Milky Way" look like? What would a project like the Sloan Digital Sky Survey reveal, if anything, of the extragalactic Universe? What would a Hubble Ultra Deep Field show? And would there be a microwave background?

Lecture Twenty-Two—Transcript
The Galaxy Web—A Relic of Primordial Sound

We're coming to the end of our six lectures on the growth of cosmic structures, and for the most part we've focused on the construction of galaxies. And occasionally, we've looked at the way galaxies gather into groups and clusters. But in this lecture, I want to step right back and take in the really big scene. How do galaxies fill the volume of the Universe? Are they spread about roughly uniformly, or randomly, or are there patterns?

It's a bit like shifting from thinking about individual people and cities and asking, "How do cities cover the Earth?" For example, there are vast areas with none (the oceans), places where there are many (around a metropolis), and lines where there are many (coastlines and along major road systems). And surprisingly, this is roughly mirrored in the distribution of galaxies. There are regions with very few (voids), dense regions with thousands (clusters), and long filaments of galaxies (the galaxy web).

So let's start close to home with a lovely movie that takes us on a trip from the Earth out to the Virgo Cluster of galaxies, about 60 million light-years away, so our average speed will be 10^{13} times the speed of light. We begin by lifting up out of the Milky Way galaxy to see its full splendor. As we back away, 2 dwarf companions pass by on the left, the Large and Small Magellanic Clouds. Remember, big galaxies like ours typically have a few smaller ones in orbit about them.

So as the Milky Way recedes, we pan around, and the other 2 large galaxies that make up the Local Group swing into view, first M33, and behind it the great Andromeda galaxy, M31. Notice these 3 big spiral galaxies are quite far apart, a necessary condition for them each to have a nice thin spiral disk. Our trip now heads away from the Local Group. We pass M81 and M82. They're close enough that M82 has been disturbed and is now a starburst galaxy. We also pass M101, and the Whirlpool galaxy, M51, whose little flyby companion has induced that lovely spiral pattern.

Our view then wheels around and shows the wider distribution of galaxies. They seem gathered roughly along a line, with a huge concentration on the left. That's the famous Virgo Cluster of galaxies, and that's where we're heading. So our trip joins the line of galaxies to the right and then moves down along it towards the main cluster. This mirrors, at hugely amplified speed, the infall of these galaxies down onto the cluster, to build it larger. The movie ends by heading straight for the dominant central galaxy, M87. And the last thing you notice is a faint blue jet on the left, a telltale signature of the 3-billion-solar-mass black hole that lives down at the galaxy's center.

Now, it's all very well, me showing a movie where we're zooming through millions of light-years, but in reality we're stuck here on Earth. How do we know what the 3-dimensional distribution of galaxies is? Actually, in principle, it's not that hard. Here's a picture of a 100° chunk of the night sky, with the Big Dipper for scale. All the foreground stars have been removed, and the image shows 2 million faint galaxies. Of course, some of those galaxies are near, and some are far, so to turn this 2-dimensional picture into a 3-dimensional one, we simply need to measure the distance to each galaxy.

Now, to measure real distances to so many galaxies would be impossible, but fortunately the Hubble law comes to our rescue. Remember, a galaxy's red shift is directly proportional to its distance, so an excellent first stab at getting distances to all those galaxies is to measure their red shifts—the greater the red shift, the further the galaxy. It's easy. So for the past 30 years, there's been a huge effort to make redshift surveys. The pioneers of this were a group at the Harvard Smithsonian Center for Astrophysics. They used a 60-inch telescope at Mount Hopkins to take spectra, one by one, of over 2000 galaxies, and ultimately, over a 10-year period in the late 1970s and '80s, they increased this number to over 18,000.

But in the last 10 years, 2 teams have now increased that number to over a million measured galaxy red shifts. Not surprisingly, they've used bigger telescopes, and also spectrographs that use optical fibers to record spectra of several hundred galaxies at one time. One team used the 150-inch Anglo-Australian telescope and its 400-fiber spectrograph to measure over 200,000 galaxy red shifts, and the other team used the Sloan Digital Sky Survey's 100-inch telescope and its 640-fiber spectrograph to measure 800,000 galaxy red shifts. These are huge projects that generate vast quantities of data and require large teams to manage. The Anglo-Australian team had 30 astronomers, and the U.S. Sloan team has 150 astronomers.

So what do these galaxy maps show us? Well, it's a little difficult to show the entire data sets, so instead, cosmologists often take a moderately thin slice through the data that corresponds to an arc-like triangular wedge, with us at the apex. So here's a typical wedge, with about 100,000 galaxies, each plotted as a single dot. Notice, we're at the apex, and the depth goes out to about 2 billion light-years, so our entire trip to the Virgo Cluster took us no further than this little circle. This is a gargantuan map. It is 1/7 the way out to the 14-billion-light-year visible limit to the Universe.

Here's our scale model from Lecture Two, where the visible Universe spans the United States. The Milky Way is a 20-yard disk somewhere near the East Coast, the Local Group spans about 0.5 miles, the Virgo Cluster is 7 miles away, and these maps span 500 miles. It's a significant chunk of all there is to see. One might view our current situation like explorers around the time of Columbus. We've made a significant start at mapping what is a finite realm. As we'll see in Lecture Thirty-Five, there are plans to measure, with somewhat less accuracy, the distance to a billion galaxies, almost out to the visible limit. This is extraordinary. Within 400 years of Columbus, no part of the Earth's surface remains unphotographed, and within 100 years of today, maybe no part of the visible Universe will remain unmapped. I'm not sure whether to feel awestruck at our capability or disappointed that Nature yields to our instruments so easily—probably a bit of both, actually.

So back to the galaxy distribution. Notice how it has an interesting pattern. It's neither purely random nor purely regular—it has a mottled look. There are actually three types of features to look for. There are large regions with very few galaxies, called "voids." Then there are long, narrow chains of galaxies, which can be sheets in 3 dimensions, and these border the voids, so the whole pattern looks like a 3-dimensional spider's web—hence the term the "cosmic web." And where the chains intersect, there are rich clusters of galaxies, which can be 1000 times denser than the voids.

This whole pattern is really quite fascinating. Why does the Universe look like this? What crafts this weblike structure? Well, by now, you won't be surprised to hear that this kind of structure emerges when gravity acts, over billions of years, on the primordial roughness that emerged from the Big Bang. The best way to demonstrate this is to shift our attention from the observational work to computer simulations that re-create the growth of large-scale structure. Now, unlike previous simulations I've shown you, which modeled one or a few galaxies, we now need to simulate regions several billion light-years across, containing millions of galaxies. So these cosmological-scale simulations are truly enormous in scope, and they push supercomputers to their limits.

This graph shows the progress in computing power applied to cosmological simulations since 1970, when only 300 gravitationally interacting objects could be followed. In 2005, that number was 10 billion. That simulation was done by the Virgo Consortium, a group of about 16 primary scientists, with about 60 collaborators from 5 countries.

Here's a region similar in size to the survey I was just showing you. It's about 2 billion light-years across, and like them, it's a thin slice, about 60 million light-years thick. The full simulation followed the separate motion of 10 billion small samples of dark matter, each one weighing 1/1000 of the mass of the Milky Way. They were initially spread out to match the lumpiness seen on the microwave background, and their simple, gravitationally induced motion was followed forward to today. In addition, gas and stars were included, using the kind of semianalytic methods I described in the last lecture. Ultimately, then, 18 million dark matter halos formed, of which 2 million contained galaxies. In fact, here's the same region, showing the distribution of galaxies.

Let's get a better idea of the 3-dimensional distribution of dark matter with this trip through the simulation. You can really sense the filaments made up of many halos of all sizes, with voids between and denser regions where they intersect. While we're zooming along here, I'll just mention that the supercomputer did 200 billion calculations per second for 28 days, marching through cosmic history at a rate of 6000 years for every second of computer time.

Okay, in this part of the trip, we're approaching a huge cluster, with thousands of dark matter halos. You can also see the thick filaments that feed more halos down onto the cluster, constantly building its size. So let's now quickly check to see how well the simulation matches the real Universe. Here's a comparison between the real galaxy surveys to the top left in blue and purple, and similar-sized regions extracted from the simulation to the lower right in red. The overall appearance of walls, voids, and clusters is remarkably similar, both visually and when you do a proper statistical analysis, so it really does seem that our picture of gravitational structure formation is at least approximately correct.

Of course, I've just shown you the final patterns, but the simulation allows us to watch how the structure got that way. So let's look at that now. Here's the initial, almost smooth distribution of dark matter, just 200 million years after the Big Bang, around the end of the dark age. You can already see areas of slightly higher and lower density, which I've marked with these shapes. Okay, let's start the movie. The redshift stretch factor is given at the top left, and as usual, cosmic expansion is not shown. We just follow the same region as time moves forward.

Notice how quickly matter moves away from the lower-density regions to the higher-density regions, enhancing the density contrast, so the web of filaments steadily grows more defined. As time passes, notice how the filaments near big clusters contain hundreds of galaxies that are falling down onto the cluster. One might even call these "rivers of galaxies." Just like normal rivers, they're flowing downhill, where in this case, downwards is towards the large mass of the cluster. And just as lakes grow as rivers feed them, so clusters grow as these filaments feed them. Well, now the movie's over. Let's just check that our initial slightly higher- and lower-density regions indeed turned into the clusters and voids. And you can see they do.

Perhaps the most bizarre aspect of all this is how such narrow filaments seem to magically appear. Why is that? Well, imagine a terrain of rolling hills. The peaks represent the highest-density regions. These highest peaks collapse first, and because their tops are roughly symmetrical, their collapse in 3 dimensions is roughly spherical. But now think about a region between 2 adjacent peaks—it takes the form of a ridge. It's the path you would hike from peak to peak. When gravity acts on a line of higher density, it tends to collapse inwards to make, in 3 dimensions, a filament. So filaments form along lines between denser regions.

Here's another movie from a somewhat smaller simulation, but it has the advantage of showing us the formation of structure in 3 dimensions by rotating the computational volume as it evolves. As before, the initial conditions are quite smooth, but as time passes, the clusters and filaments form. At this point, I'd like to reintroduce the roughness spectrum from Lecture Seventeen. Remember that all this complex structure can be represented as the sum of many, many waves, each of different wavelength and strength. It's a roughness spectrum.

Here's the cosmic roughness spectrum for today's distribution of galaxies. On the left are large scales inferred from the microwave background, and on the right are smaller scales measured from the galaxy redshift surveys. It has four main features: It rises on the left, there is a peak, it falls on the right, and there are some small wiggles on the right. The shape means that, moving from right to left, you can find ever-larger structures up to a maximum of about 200 million light-years, but beyond that the Universe is progressively smoother on larger scales. Remember, it's homogeneous on large scales.

So what about those wiggles? What are they from? Well, if you think back to Lecture Fifteen on Big Bang acoustics, those little wiggles are the faint remains of the acoustic harmonics in the young Universe, before the fog cleared, before 400,000 years. Here's the sound spectrum from the microwave background, with its fundamental and harmonic peaks, and the odd compression harmonics survive through to today's roughness spectrum as those little wiggles. They're sufficiently weak that your eye can't see them as patterns in the galaxy maps, but a proper analysis shows that they're there as wiggles in the galaxy roughness spectrum. Their technical name is "baryon acoustic oscillations." This is a truly wonderful notion. Primordial sound is still with us, writ large across the sky in the patterns of galaxies.

Well, now, let's come closer to home and add something that's been implicitly present all along—the flow of galaxies, including our own. Although we've seen simulations showing galaxies moving from voids to filaments to clusters, can we actually detect this movement? Can we see the galaxy flow pattern? The answer is "yes," though it's not very easy. Here's the method. Let's think back to how galaxy maps are made, using red shifts. With pure cosmic expansion, a galaxy's red shift should give us its distance exactly—that's just the Hubble law. But if the galaxy has an additional motion because it's been pulled by nearby galaxies, then the measured red shift includes this additional part.

Now, usually, in the maps, that's just thought of as an error on the distance. But now imagine you were able to measure the actual distance to the galaxy, independently, using one of the methods from Lecture Six—let's say the Tully-Fisher method. Now you know the part of the red shift linked to distance, so you can subtract it off the measured red shift, and bingo, the difference gives you the galaxy's gravitational motion. So combining redshift pie maps with accurate distances allows you to construct the galaxy flow pattern.

Now, let's be clear here: We are not talking about the global cosmic expansion, the "Hubble flow," as it's called—that's a given. We're talking about motion in addition to the Hubble flow that's arisen over time, as galaxies pull on one another. Here's an analysis of the gravitational velocities in the local galactic neighborhood, where "local" here means out to about 300 million light-years, about 10 times smaller than the pie maps I showed you earlier. It turns out that much of the action happens roughly within a giant plane called the "supergalactic plane." So we won't miss too much if we just show this slice of the local Universe on an x, y graph, omitting the z dimension.

Here's the distribution of galaxies, with darker shading for more galaxies. We're at the center, in a relatively low-density region with the huge Pisces-Perseus supercluster of galaxies to the right, and the Centaurus and Great Attractor superclusters to the left. Now, here's the measured galaxy flow pattern, using little arrows to indicate the speed and direction of the flow at each point. And to keep the figures clean, I've shown a model fit to the data, rather than the data itself.

The results are quite impressive. Look how the flow always carries galaxies towards the denser regions and away from the voids, just as we expect. Galaxies are indeed falling downhill from voids to clusters. Actually, the speed at which they're falling is much higher than you would expect if they were only being pulled by the galaxies. As always, dark matter is dominating all of this gravitational motion, and the galaxies are really just along for the ride, being pulled by the dark matter.

Okay, back to the Milky Way. Notice it's caught in a tug-of-war between those 2 huge collections of galaxies, the Pisces-Perseus supercluster on the right and the slightly closer Great Attractor on the left. The Great Attractor is winning the battle, hence its name, and the local region, including the Milky Way, is falling towards it at roughly 500 kilometers per second. Now, don't be confused here— we're still expanding away from the Great Attractor at 4500 kilometers per second, but we would be expanding at 5000 kilometers per second if it weren't there. It has slowed us down by 500 kilometers per second. So it's only relative to the Hubble flow that we are, in fact, falling at 500 kilometers per second.

Let's now quickly zoom into this central square to look more closely at our own suburb, before we back out to look at an even bigger scene. Here's the Local Group, with the Milky Way and M31, sitting in a small filament that extends up to meet the much larger filament, with the Virgo Cluster at its center. That's where our flight simulator movie took us at the beginning of this lecture. There also seems to be a void behind the Virgo Cluster, and there are other smaller clusters to the right and below. The arrows show the flow pattern, which is, once again, away from the voids and towards the denser regions. And as you might expect, the Local Group and Milky Way galaxy are falling towards Virgo, moving at about 300 kilometers per second, relative to the Hubble flow, and of course, this entire region is moving off to the left at 500 kilometers per second, falling towards the Great Attractor.

So let's finally step back to look at the surrounding 800 million light-years, about the central fifth of those earlier pie maps. You can see the 2 previous regions with their concentrations of galaxies, but now you can see several much larger concentrations—the dense Coma Cluster above, and the huge Shapley concentration, about 500 million light-years away, and containing tens of thousands of galaxies. Looking at the velocity field, it seems that although the Shapley cluster concentration strongly affects its local region, at 500 million light-years away, it's too far off to affect us directly. We primarily feel the pull of the Virgo Cluster and the Great Attractor.

Now, let me end this lecture by looking at these large-scale flow patterns in a slightly different way. Remember, you can view all of structure formation as a hierarchy of collapse and merger. Early on, small, high-density regions turn around and collapse. Later on, larger, high-density regions do the same thing: expand, slow, turn around, and collapse. And they include many of those earlier, smaller objects in their collapse. Well, these large-scale flow patterns that we've just been looking at are really just the latest stage in this game. We have huge regions that are just beginning to slow their expansion, pulled back by their higher density.

Now, here is an interesting question: Will they ever finally halt, turn around, and collapse to form super-duper clusters of galaxies? The answer is "no," they won't, and here's why: Dark energy has begun to affect cosmic expansion. It's driving acceleration, and this significantly affects the future growth of structures. Basically, the growth is shut down and frozen in its current pattern.

This figure shows a simulation of a billion-light-year region that is designed to match our own neighborhood. So we're at the center—here's the Virgo Cluster—and you can see the same galaxy patterns I've just been showing you. Now let's jump 14 billion years into the future. That billion-light-year circle now appears smaller because, as usual, we're not showing the expansion. Okay, let's jump another 14 billion years into the future. Notice how the structure just isn't changing. It's been frozen by the accelerating expansion. Of course, by this time, the Milky Way and Andromeda merged long since, but we never fell into Virgo, and Virgo never grew much larger.

Now let's jump 100 billion years in the future. The billion-light-year circle is now a dot at the center, and a new, red, dashed circle has appeared. That's the distance at which the expansion speed is the speed of light. It's called an "event horizon." Now, as you might imagine, you can't see anything outside this circle because the light never gets to you. Ultimately, even the Virgo Cluster will be outside the event horizon, and the Universe will appear empty, except for our own local ball of stars, what was once the Local Group.

Let me very briefly review the main points of this lecture. Rather than focus on individual galaxies, we stepped back and looked at the overall pattern of their distribution. This pattern has become visible, using huge surveys of over a million galaxy red shifts, which span several billion light-years. We see a weblike pattern, with voids, filaments, sheets, and clusters. All these features are nicely reproduced using giant computer simulations that follow the motion of literally billions of objects that represent dark matter, starting from the slight roughness in the early Universe. Using these simulations, we can see how the patterns steadily become clearer, as gravity pulls galaxies from lower-density regions to higher-density regions, and we can even see the remains of the acoustic harmonics left over from the sound waves in the million-year-old Universe.

We also saw how combining distance measurements with redshift surveys allow us to map out the galaxy flow patterns, superimposed on the Hubble expansion. It turns out locally, we're falling towards the Virgo Cluster, but we and the Virgo Cluster are both falling towards the Great Attractor. Finally, we learned that this ongoing growth of structure will soon shut down, as the accelerating expansion caused by dark energy freezes the pattern of galaxies more or less in its current form.

Well, our lecture course is a little over halfway through, so I hope you're ready for a change of topic. We have several wonderful themes coming up, starting in the next lecture with the story of how all the atoms that make up our world, including you and me, were first made.

Lecture Twenty-Three
Atom Factories—Stellar Interiors

Scope:

The next three lectures place us, and the atoms from which we're made, into the cosmological story. After reviewing the nature and structure of atoms and their nuclei, we'll look at the factories that make them: stars. Deep in the heart of a star the conditions are ripe for thermonuclear fusion—the building of heavier nuclei from lighter ones. Toward the end of a star's life, the newly made nuclei come to the surface, where they gain their electrons to become full atoms. Depending on how the star dies, these atoms are ejected calmly or explosively into space, where they may drift for several billion years before finding their way into a denser gas cloud, and from there into new stars and planets. This is the story of the atoms that made the Earth and created the biosphere, and even those that now help you read this sentence.

Outline

I. Does modern cosmology tell the story of our own origins? Yes, it must! We're no less a part of the Universe than stars or galaxies. The next three lectures look at the creation of the atoms from which we, and everything around us, are made.

II. Atoms are incredibly tiny and numerous: If they were the size of marbles, your hand would be as big as the Earth.

    A. There are about 100 kinds of atoms forming an ordered sequence. Each makes a particular chemical element, for example, carbon (6), gold (79), or uranium (92).

        1. The sequence is usually shown wrapped around in a "periodic table."

        2. Atoms stick together to make molecules, which can in turn make cells, organs, and even people.

    B. Atoms are themselves composite.

        1. Lightweight, negatively charged electrons orbit a heavy, positively charged nucleus.

        2. Nuclei are 100,000 times smaller than atoms. If you remove the space within atoms, the human race fits in a sugar cube.

3. Nuclei contain protons and neutrons stuck together by a "strong force."

C. The "nuclear binding energy curve" is an extremely important graph.
   1. It plots, for the sequence of atoms, the glue energy holding each nucleus together.
   2. The curve starts high with hydrogen, drops to a minimum at iron, and then slowly rises to uranium.
   3. The shape reflects the competition between the short-range attractive strong force and the long-range repulsive electric force of the protons.

D. Reactions that move down the curve release energy.
   1. These include fusion reactions, which combine light nuclei, and fission reactions, which break apart heavy nuclei.
   2. These reactions release about a million times more energy than chemical reactions (per kilogram).
   3. Fusion reactions are difficult to start because nuclei repel each other strongly. However, at high enough temperature, nuclei can collide and stick in "thermonuclear fusion" reactions.

III. Thermonuclear fusion provides our link to the stars. Ultimately, you can think of stars as "atom factories."

A. The huge weight of a star makes a hot, dense core.

B. The core of a star is like a furnace.
   1. It "burns" lightweight fuels into heavier ones. This releases energy, which heats the core and keeps the reactions going.
   2. Heat also moves up to the surface, which glows brightly because it is hot—the star shines!

C. The ash from one reaction becomes the fuel for another. For example, hydrogen burns to helium, which burns to carbon, and so on.
   1. The sequence can't get past iron, because it sits at the bottom of the binding energy curve.
   2. The sequence needs ever-higher temperature, because nuclei with more protons repel more strongly.
   3. How far along this sequence a star gets depends on its mass. Higher mass stars can make heavier elements.

IV. How do we get the freshly made nuclei out of the star's core and into you and me?

    A. It starts with "convective dredge-up." Nuclei are brought to the surface in huge, hot currents of gas.

    B. On their way up, at lower temperature, the nuclei acquire their quota of electrons and become atoms. The atom factory has its "conveyor belt."

    C. These atoms are finally ejected into space when the star "dies," meaning when the fuel runs out and the nuclear reactions cease.

V. There are several ways in which stars die. Here are three.

    A. When low-mass stars (similar to the Sun) run out of fuel, their core contracts, and vigorous fusion just above the core generates lots of energy that makes the star swell up to become a "red giant."

        1. Being so large, the star's gravity can't easily hold onto its surface, which gets pushed off as a wind to make a beautiful nebula.

        2. Ultimately this gas, with its new atoms, drifts off into space.

        3. The remaining core of carbon and oxygen is extremely dense. These "white dwarf stars" normally just cool and fade over billions of years.

    B. However, if the white dwarf is in a binary star system, it can be rekindled.

        1. When the second star swells to become a red giant, it transfers gas onto the white dwarf, which gets compressed further and heated.

        2. At a critical moment, the carbon and oxygen undergo thermonuclear detonation, and the whole star explodes as a supernova.

        3. Many new elements, especially around iron, are hurled into space.

    C. More massive stars die in a different kind of supernova explosion.

        1. As their fuel runs out, they swell to become "red supergiant stars."

2. Deep in the core, heavier elements fuse in a sequence of nested shells, the innermost being iron—the ultimate nuclear ash.

3. At a critical moment, the pressure supporting this iron core fails, and it collapses, triggering a monumental explosion.

4. The explosion makes elements up to uranium and hurls them out into space, where in time they join the tenuous gas between the stars—the "interstellar medium."

VI. This is a new stage in the lives of our newly made atoms.

    A. The interstellar gas is very thin: maybe 1 atom per cubic centimeter.

    B. Atoms of heavy elements also make tiny smog particles that, overall, comprise 2% of the interstellar material. In a sense, the galaxy's atmosphere has been very badly "polluted" by all of those atom factories.

    C. The atoms in you and me probably drifted around in the interstellar medium for 1 or 2 billion years before joining a denser cloud.

    D. Within such clouds, small pockets collapse to form stars, and around these one finds disks of dust and gas, which in turn form planets.

VII. In the case of the Earth, about 10^{50} atoms ended up in a spherical ball, with a barren, cratered surface heaving with volcanism.

    A. Over 4.5 billion years, an extraordinary transformation took place, enabled by atoms' amazing ability to combine in complex ways.

    B. At glacial pace, a selection process took us from molecules to life.

    C. Ultimately, with unimaginable dexterity, instructions inherited from long lines of ancestors coax billions of atoms into gargantuan structures.

    D. Perhaps most remarkable of all, some of those atoms are, right now, comprehending this sentence and, in a sense, thinking about themselves.

Theme V: The Origin of the Elements

Suggested Reading:

Hawley and Holcomb, *Foundations of Modern Cosmology*, chap. 5.

Ferris, *The Whole Shebang*, chap. 4.

Silk, *The Big Bang*, chap. 16.

Rees, *Just Six Numbers*, chap. 4.

Questions to Consider:

1. What is the nuclear binding energy curve (shown in the lecture in a graph), and why does it have the shape that it does? Use this graph to explain why fusion of light nuclei and fission of heavy nuclei both release energy. What is it about the iron nucleus that makes it sit at the bottom of the curve? What astrophysical significance does this nucleus have?

2. Why do thermonuclear fusion reactions require such high temperatures? Why do reactions between heavier and heavier nuclei need higher and higher temperatures? Why are low-mass stars unable to "burn" these heavier nuclei while high-mass stars can?

3. Describe the buildup of chemical elements as a galaxy ages. Which stars make which elements? How do stars return new elements to the galaxy to maintain the cycle? Compare how a Sun-type star dies with how a 15-solar-mass star dies. How do each of these eject their newly formed atoms?

4. Let's compare, kilogram for kilogram, the Sun's nuclear furnace to your metabolic "furnace." How many watts of power per kilogram does each furnace generate? For the Sun, use $L_{Sun} = 4 \times 10^{26}$ watts and $M_{Sun} = 2 \times 10^{30}$ kg; and for you, use $L_{you} = 100$ watts and $M_{you} = 80$ kg. Amazingly, your digestion is about 10,000 times more powerful per kilogram! How can this be, given the Sun's enormous power and the energy richness of its fuel? Hint: Although a star made from 10^{28} people (2×10^{30} kg) would be 10,000 times brighter than the Sun, it would burn out in a week (no eating allowed!), whereas the Sun lasts 10 billion years.

Atom Factories—Stellar Interiors

In the last few lectures, I've told the story of how huge structures have grown in the Universe as it has aged: the birth of the first stars, the maturation of galaxies, and the development of the galaxy web. Our focus couldn't have been further from our cozy home here on Earth. But surely an important part of any creation story is how our own very local world came into being, including, and perhaps most importantly, how we ourselves came to be here. What does modern cosmology have to say about this? Does it deal with such mundane things as this podium or my pen or even my hand? Does it tell their story?

The answer is "yes," of course. It must do. We are no less a part of the Universe than any star or galaxy, and we can only have arrived here via a cosmological route. So it's this route I want to explore in the next three lectures, and I hope by the end of them, you will look differently at everything around you, and you may even feel your own cosmic origins. Let's start by carefully looking at my hand. What's it made from? As you probably know, if we zoom ever deeper down in scale, we first pass skin, then cells, then organelles within those cells, then giant molecules, and finally, about 8 factors of 10 in magnification, we arrive at an ocean of tiny jostling spheres. These are the atoms from which everything is made, and they are famously tiny compared to us.

If this marble were the size of an atom, then my hand would be the size of the Earth. Imagine fingers 5000 miles long, 1000 miles across, packed with marbles. Atoms are incredibly tiny, and they're incredibly numerous. Now, as you probably know, there are about 100 different kinds of atoms, which differ in weight and size and how they stick to each other. The most famous way to organize them all is in the periodic table of the elements. That's something I suspect you learned about in high school. Each element is made from atoms of just one kind, and they're all arranged in a sequence, from 1 to about 100, that wraps around so that elements with similar chemical properties all line up in columns.

Here are just 4 examples. Lithium and carbon are numbers 3 and 6. Gold and uranium are numbers 79 and 92. Of course, it's the amazing ability of atoms to stick together as molecules that makes atomic matter so creative. In the same way that a few dozen letters in an alphabet can make millions of words that can be arranged into sentences and books and libraries, so a few dozen kinds of atoms can make millions of molecules that can be arranged into cells and organs and people. Here are some models of simple molecules: water, ethanol, the neurotransmitter serotonin, and a tiny fragment of DNA.

Now, since our aim is to learn how atoms were made, we need to get beyond this simple cartoon of atoms as shiny, colored spheres, so here's a slightly better diagram of a beryllium atom. That's number 4 in the series. There are 4 lightweight, negatively charged electrons orbiting around a heavy, positively charged nucleus. It's the electric force of attraction that keeps the electrons in orbit, rather like the Sun's gravity keeps the planets in orbit. Now, the atom's nucleus appears way too big in this diagram. It's actually 100,000 times smaller than the atom. A fly in the middle of a football stadium is a good match to a nucleus in the middle of an atom. So if you removed the space from all the atoms in all the people of the world, we would all fit into this sugar cube. You see, relative to nuclei, atoms are very large, very empty things.

Now, despite their amazing tininess, we actually know a great deal about atomic nuclei. They contain 2 kinds of particles, protons and neutrons, and it's the number of protons that defines the element: 1 for hydrogen, 2 for helium, 79 for gold, and so on. Usually, there are slightly more neutrons than protons, so for example, gold has 79 protons and 118 neutrons. Now, the neutrons and protons are held together very tightly by a second kind of force called the "strong force," but unlike the electric force, it has a very short range. You can think of it a bit like Velcro—close enough, it sticks, but a little further, nothing. Now, this attractive Velcro force is crucial because without it, the positively charged protons, which repel each other, would just blow the nucleus apart.

Now, at this point, I want to introduce one of the most important graphs in astrophysics. It's called the "nuclear binding energy curve." Along the bottom, from left to right, we have all the elements from 1 to 92, and up the y-axis, we have the energy that holds all the protons and neutrons together. It's called the "binding energy." It's the energy you would need to pull the protons and neutrons apart. The lower a nucleus sits on this curve, the more glue energy there is, the more tightly bound it is. Notice that the curve starts very high with hydrogen, drops to a minimum, and then slowly rises again. Now, the reason it has this shape is because there are both attractive and repulsive forces at work inside the nucleus.

For lighter nuclei, with not too many protons, the proton repulsion isn't that much, so the strong attractive Velcro force dominates, and makes nice, tightly bound, small nuclei. But as we move to the right, the proton repulsion begins to build up, and competes with the attractive Velcro force, so the curve begins to rise. And really big nuclei, like uranium, are fairly close to just blowing themselves apart. The element that sits at the bottom of this curve is iron, number 26. Its nucleus is the most tightly bound.

Okay, well, now imagine the following: Let's take 2 light nuclei and fuse them to make a single heavier nucleus. For example, let's fuse hydrogen to make helium. We move to the right and downwards on this graph. Since lower on the graph is more tightly bound, then energy is released. Think of rolling a ball downhill. So fusion reactions that combine light nuclei together release energy. It's these kinds of reactions that power stars and also hydrogen bombs. Now, look to the right. If we break apart a heavy nucleus like uranium, we move to the left on the curve, which is also downhill, so energy is also released. These are fission reactions, and they power nuclear reactors and also atomic bombs.

What's not so apparent from this particular graph is that the actual amount of energy released in nuclear reactions is about a million times greater than in chemical reactions. For example, fusing a kilogram of hydrogen to make helium releases about a million times more energy than burning a kilogram of coal to make carbon dioxide. Of course, burning coal is easy, but fusing nuclei is difficult.

Why is that? It's because nuclei are all positively charged, so they repel each other like crazy, and it's very difficult to get them together. One way to do this is to rely on very high temperatures to hurl nuclei together, such that they can actually collide and stick. The speeds need to be very high, typically a few thousand kilometers per second, corresponding to temperatures of millions, or even billions, of degrees. These kinds of reactions are called "thermonuclear fusion" because they rely on high temperatures to get them going.

Now, it's these thermonuclear reactions that provide the crucial link to the nature of stars. You see, the reason stars, like the Sun, are hot, luminous balls of gas is that they are busy converting light nuclei into heavier ones, and the energy released keeps the star hot and shining. Ultimately, then, new elements are built up inside stars, and that overall process is called "nucleosynthesis," the synthesis of atomic nuclei. The bottom line is stars are atom factories. So let's look in a bit more detail now at how all of this actually happens.

Here's a star. It happens to be the Sun, but it's going to represent stars in general. Let's cut into it to see its interior. Stars, of course, are huge things, with huge quantities of matter, whose enormous weight compresses the interior to make an extremely dense, hot core. At the center of the Sun, for example, the gas is 10 times denser than lead, and 14 million K. Now, these conditions are ripe for thermonuclear fusion. In a very real sense, the core of a star is like a furnace, and the rest of the star is like an insulating jacket. Within the furnace, a fuel of lightweight nuclei is being burned in the nuclear fusion sense to make an ash of heavier nuclei, and that burning gives off energy, heat.

Just like a normal furnace, the heat generated from the burning does two things. It keeps the core hot so that the burning can continue, and it also moves out through the insulating jacket, and ultimately it escapes into the colder environment. Now, in the case of the Sun, the energy makes its way out of the star in a couple of ways: X-rays slowly diffuse up through the foggy interior, and huge convection currents carry hot gas from below up to the surface. The surface is really the part of the star where the gas turns transparent so that the light from the hot, glowing gas is set free and streams out into the darkness of space. So when you next look at those points of light in the night sky, try to think of them as glowing surfaces of distant atom factories. They're a direct indication that new atomic nuclei are being made.

There is a key property of stellar furnaces that you won't find in normal human furnaces. The nuclear ash from one thermonuclear reaction can become the nuclear fuel for another thermonuclear reaction. Let's go back to our binding energy curve. The first thermonuclear reaction fuses hydrogen to make helium, but when all the hydrogen fuel runs out, helium nuclei can fuse to make carbon and oxygen. And when the helium runs out, the carbon and oxygen nuclei can begin to combine to make neon and sodium, and so on and so on. The nuclear ash from one reaction becomes the fuel for the next, and this sequence steadily creates heavier and heavier elements.

But there are two catches to this simple story. First, it only works up to iron, which sits at the bottom of the binding energy curve. It's the ultimate nuclear ash because to go beyond it is uphill, which of course absorbs energy, not releases it. The second catch is that these later reactions need much higher furnace temperatures for the very simple reason that heavier nuclei have more protons, so they repel each other even more forcefully, so you need even higher temperatures to get the nuclei together. For example, hydrogen fuses at 15 million K, helium fuses at 100 million K, carbon fuses at 600 million, and oxygen fuses at 1.5 billion K.

Now, whether or not a star can actually achieve such high furnace temperatures depends on the mass of the star. You see, when one fuel runs out, the core of the star—that's the furnace—begins to contract under its own weight, and this compresses the core and heats it up. How hot the core gets depends on how massive it is. Relatively low-mass stars like the Sun can never ignite carbon, while intermediate-mass stars can never ignite neon. It's only massive stars that can fuse elements all the way up to iron. Now, I'm going to talk about these reactions in more detail in the next lecture, including the all-important processes that make elements beyond iron, but for now, let's just recognize that the furnaces that reside deep inside stars can generate the whole sequence of atomic nuclei that we find in the periodic table of the elements.

It's all very well, making atomic nuclei deep inside stars, but here they are in you and me. How did they get from out there to in here? Let me trace that remarkable journey for you now. For many atoms, the process starts with what's called "convective dredge-up." They are brought up to the surface in huge, hot currents of gas. Of course, higher up in the star, the temperature is lower, and so one by one, negatively charged electrons get caught and attached to the positively charged nuclei, and by the time the atom arrives at the surface, it's all but one or two of its final quota of electrons. So while nuclei are built in the star's core, atoms are assembled near the star's surface. The factory has, if you like, its conveyor belt.

But we still need to get the atoms out of the star and into you and me, and that takes quite some doing. Ultimately, what happens is that the atoms are ejected into space when the star dies, and I'm using the term "die" more or less as you would for a furnace, when the fuel finally runs out and the thermonuclear reactions cease. Now, there are actually several ways in which stars die, and I'm going to mention three of them. So let's start with stars like the Sun, what astronomers think of as relatively low-mass stars.

About 5 billion years from now, when the Sun's fuel begins to run out, its core will begin to contract, and fusion reactions just outside the core begin to burn extremely vigorously and create lots and lots of energy, and this makes the Sun swell up enormously. Here's the Sun today, sitting at the center of the orbits of Mercury, Venus, and Earth. But as its furnace begins to die, the Sun will expand to engulf the orbits of Mercury and Venus, and almost touch the Earth. It becomes what astronomers call a "red giant star." Needless to say, at this time, the Earth's surface will return to an utterly barren and lifeless state.

Now, with its atmosphere so extended, a star's gravity can't easily hold onto the outer layers, and as more and more energy comes up from below, it actually pushes off the star's outer atmosphere. And over a few tens of thousands of years, the outer layers are blown away in a strong wind, and a beautiful nebula is created. The atoms in these nebulae steadily move out to join the rest of the material between the stars, and I'll return to pick up that story in just a minute.

But first, notice that in the middle of many of these nebulae, you can see the exposed, hot, dead core. These hot cores are called "white dwarf stars." They're made of highly compressed carbon and oxygen, not much bigger than the Earth, but with about as much mass as the Sun, so they're incredibly dense. Now, normally, over billions of years, they would just gradually cool down and fade away, but in certain special circumstances, they can be rekindled in a very dramatic way as a supernova explosion. This can happen when the white dwarf happens to be in a binary star system, with a second companion star orbiting nearby. When the second star begins to exhaust its hydrogen and swells up, some of its outer atmosphere falls over onto the white dwarf. As mass is transferred, the white dwarf gets compressed and heated, and there is a critical moment when its temperature is sufficiently high that the carbon and oxygen nuclei undergo thermonuclear detonation and the entire star explodes.

We actually met this kind of supernova back in Lecture Six, because they provide a fabulous way to measure the distance to very remote galaxies. But in the current context, we're more interested in the fact that thermonuclear detonation not only creates a whole set of heavy elements, mainly around iron in the periodic table, but it also hurls them out into space at thousands of kilometers per second. Now, there is, in fact, another type of supernova that comes from the death of a much more massive star, whose furnace can actually ignite elements well beyond carbon. This time, when the fuel begins to run out, these massive stars swell up even larger than red giants to become what are called "red supergiant stars," roughly the size of Jupiter's orbit.

But way down in the center sits an amazingly dense hot core. See, with such high temperature and density, elements heavier than carbon start fusing so readily that they form a set of nested concentric shells—many furnaces burning simultaneously, each receiving fuel from the shell above and dumping its nuclear ash to the shell below. Right at the heart of this is a dense sphere of iron nuclei, the ultimate nuclear ash that cannot yield any more fusion energy. It's about the size of the Earth, but with more than the mass of the Sun.

As time passes, this iron core steadily grows in mass, and in one of the most remarkable events in astrophysics, the pressure supporting this iron core snaps, and in a fraction of a second, it collapses and triggers a monumental explosion. Just to give you an idea of the energy released, if we orbited such a star and were subjected to its explosion, each square foot of the Earth's surface would be hit by the energy equivalent of 1000 times the global total stockpile of nuclear weapons. That's each square foot. These core-collapse supernovae are prodigious astrophysical bombs, but in the current context, they're also remarkably creative. They make elements past iron, all the way up to uranium, and furthermore, they hurl these new elements out into the interstellar medium.

The spreading guts of these star explosions are called a "supernova remnant." In that animation, the remnant was designed to look like the Crab Nebula, whose star exploded almost 1000 years ago. Here are 2 other young remnants that are about 400 years old, and they've now expanded to be just a few light-years across. And they're found to contain almost pure heavy elements, with little or no hydrogen or helium, which is a pretty strong indication that these elements were indeed made inside these stars.

Now, after many thousands of years, like these older supernova remnants, the gas has expanded to cover several hundred light-years, and in time, it gradually disperses and joins what astronomers call the "interstellar medium," the tenuous gas that exists between the stars. This is the next stage in the lives of our atoms, so let's take a closer look at it now. You can get a reasonable impression of the interstellar medium in this lovely wide-angle image of the Milky Way. All that patchiness is gas and dust between the stars. The gas is exceedingly thin, maybe one atom per cubic centimeter. That's equivalent to one person per continent for humans.

In addition to single atoms, there are also groups of maybe a million atoms stuck together in tiny grains that astronomers call "dust," though you can think of them as very fine smog particles. In fact, those dark patches you can see across the Milky Way are caused by this interstellar smog blocking the light from more distant stars. You see, you can view the interstellar medium as the galaxy's highly polluted atmosphere, polluted by all the heavy elements and smog particles coming from all those atom factories, the stars. They really are filthy furnaces. After 14 billion years of pollution, the Milky Way's atmosphere contains about 2% heavy elements. If you compress the gas to match our own atmosphere's density, you could not see your hand in front of your face—it is that polluted.

Of course, one can choose a more positive metaphor and say that the interstellar medium is "enriched by heavy elements." After all, they're an essential ingredient of building planets and people. So let's follow our story further. How do these atoms continue their journey into the Earth and into you and me? Well, our atoms probably drifted around in the interstellar medium for at least 1 or 2 billion years, subject to the sort of general mixing of galactic rotation. And in time, at least some of the atoms find themselves in denser clouds, like this. When gas is gathered into denser regions, gravity begins to take over, and in small pockets it can cause a headlong collapse that ends in the formation of a star and its planets. Basically, these utterly beautiful sights are where new stars are born.

Here's a lovely example: the great nebula in Orion. If we zoom into the region with the Hubble telescope's sharp vision, we can find new stars forming all over the place. Perhaps a few thousand will ultimately form, but more important for our current story is that these new stars are surrounded by a flattened disk of gas and dust, and it's within this disk that planets form. As this artist's painting suggests, conditions in the disk allow gentle collisions to build larger structures, so atoms stick to form grains, grains form pebbles, pebbles form boulders, and at some point gravity takes over and attraction onto the largest boulders builds first embryonic planets and eventually full-scale planets.

In the case of the Earth, about 10^{50} atoms ended up in a spherical ball, looking maybe a little like this, with a barren, crater-pocked, sterile surface, heaving with volcanism. In one of the most remarkable transformations of all, over a period of 4.5 billion years, this young, sterile planet gradually turned into our current world, with warm oceans filled with fish, and continents teeming with plants and animals. That transformation has its own wonderful story, and is, of course, way beyond the scope of these lectures. It has its roots in the amazing ability of atoms to bind together in a myriad of complex ways. At glacial pace, an extraordinary selection process begins to leave its mark. Molecules that are stable and can duplicate gradually become more common, and this principle continues in a hierarchy of structures that ultimately leads to life. Natural selection and the passage of time are superb craftsmen, with unimaginable dexterity, instructions inherited from long lines of ancestors, coaxing billions of atoms into huge structures.

After thousands of millions of years, it's even possible for atoms to arrange themselves into plants, animals, and people. And that brings this lecture full cycle. We set out to track our own matter's history, to find its location in the cosmological story, and what a story it's turned out to be! Our atoms were forged billions of years ago, deep in the fiery furnaces of ancient stars, long since dead. When those stars died, the newborn atoms were launched out into the cold emptiness of interstellar space, where they drifted for several billion years. In time, they joined a dense, collapsing cloud and became incorporated into the Sun and planets. A select few ended up on the Earth's surface and spent time moving from plant to animal to soil, and back again. And for just a brief moment in their long, long lives, billions have gathered to make your body, and some of these are, right now, comprehending this sentence. And in a sense, they're actually thinking about themselves.

Lecture Twenty-Four
Understanding Element Abundances

Scope:

How do we know whether our story of atom genesis from the last lecture is true? The key test is to measure and explain the *abundance* of each element. The abundances of astronomical objects are remarkably similar: a steady decline in abundance for heavier elements, with several peaks along the way. We'll explore how the entire element series is made and explain the various features on the abundance graph. We'll learn that making helium and carbon in stars is extremely difficult but beyond that, elements up to iron can be made relatively easily. Iron is a watershed, lying at the bottom of the binding energy curve; it is the end point of thermonuclear fusion. To go beyond, we need neutrons to push nuclei further up the sequence, either slowly in the S-process or rapidly in the R-process. Ultimately, in one of the 20th century's greatest scientific triumphs, it's now possible to explain the entire abundance graph almost perfectly.

Outline

I. In the last lecture, I simply told you the story of how our atoms were made. How do we know it's true? By looking at the *relative abundances* of the elements. That's our theme for this lecture.

   A. The light from stars reveals their composition, because their atoms emit and absorb specific sets of colors. These spectral "bar codes" tell us which elements are present.

   B. Measuring the amount of each element is more difficult. In 1925, Cecilia Payne figured out how to do this, and she discovered that stars contain mainly hydrogen and helium.

   C. Measuring abundances is now a mature subject and has yielded three primary results.

      1. Almost everything contains three-fourths hydrogen, one-fourth helium, and 0%–4% heavier elements. (Earth is an exception: Its weak gravity can't retain hydrogen or helium.)

      2. The heavy-element fraction varies in ways we'll talk about in the next lecture.

    3. The relative abundances *within* the heavy elements are very similar in all objects.

II. A graph of the element abundances is extremely important. The *x*-axis is the element series (hydrogen to uranium); the *y*-axis is the relative abundance.

    A. Overall, its shape is a steady decline, with a number of peaks.

        1. Hydrogen and helium are by far the most abundant.

        2. There is a huge drop to the "light elements" (Li, Be, B).

        3. Then comes the "alpha peak" (C, O, Ne, Mg).

        4. Then the "iron peak" (Co, Fe, Ni).

        5. Then there's a gradual decline to extremely rare elements, with 3 pairs of smaller peaks, labeled "R" and "S."

    B. Our aim is to explain this curve. As we'll see, knowledge of atomic nuclei and star furnaces can in fact explain it very well, including all its peaks.

III. Let's now consider how the sequence of elements was made, starting with the first step from hydrogen to helium.

    A. This requires 2 protons to combine to make deuterium.

        1. This is an extremely difficult reaction because it involves the weak nuclear force. Its slowness is the main reason stars live so long.

        2. Once made, deuterium is relatively easily converted to helium.

    B. Going beyond helium is also very difficult.

        1. Simple 2-particle reactions involving helium won't work, because they all make unstable nuclei.

        2. In the early 1950s, Ed Salpeter and Fred Hoyle solved the problem.

        3. When a star's helium core becomes highly compressed and hot, 3 helium nuclei collide almost simultaneously to make carbon-12. This is called the "triple alpha reaction" (another name for a helium nucleus is "alpha particle").

    C. The way to make heavier elements was described in 1957 in a classic paper by Margaret and Geoff Burbidge, Willi Fowler, and Fred Hoyle. Much of the story I'm about to tell you has its roots in this paper.

D. Going beyond carbon is relatively easy.
  1. Helium nuclei can be added in a simple sequence: carbon-12 to oxygen-16 to neon-20 to magnesium-24 to silicon-28 to sulphur-32.
  2. These are called the "alpha elements." They're quite common and make up the alpha peak on the abundance graph.
  3. The less-common elements in between (e.g., fluorine, sodium) are made in more complex reactions at even higher temperatures.
  4. Recall, heavier nuclei repel each other more strongly, so all these reactions need very high temperatures: 100 million to a billion K.

E. Beyond the alpha peak is the iron peak. Why are these elements common?
  1. A quick look at the binding energy curve shows why: Iron sits at the bottom. Fusion reactions release energy up to iron but not beyond.
  2. Without going into details, fusion tends to create large quantities of elements 24–28, which form the "iron peak" on the abundance graph.

IV. To go beyond iron, we need a mechanism other than fusion. This is where neutrons are important.

  A. In certain stars, free neutrons can be present; and with no charge, they can enter nuclei very easily.
    1. Adding neutrons to nuclei makes them unstable, and they decay by spitting out an electron. In effect, a neutron is converted to a proton.
    2. In this way, one can steadily move up the sequence of elements to higher proton numbers.

  B. As heavier elements are made, a fascinating phenomenon occurs. At a few particular points along the sequence, nuclei build up and have much higher abundance. Why is this?
    1. Protons and neutrons orbit within the nucleus in shells, just like electrons orbit within atoms.
    2. In atoms, when an electron shell is filled, that atom is very stable—it's a noble gas. This is also true for nuclei. When a neutron shell is filled, that nucleus is very stable.

3. The particular numbers when this happens—50, 82, and 126—are called "magic numbers," and the nuclei are called "magic nuclei."
4. As heavier nuclei are made, they collect at the magic neutron numbers, around zirconium, barium, and lead.
5. These make one set of the peaks on our abundance graph. They're called the "S-process peaks," where "S" stands for "slow neutron addition."

C. The "R" in "R-process peaks" stands for "rapid neutron addition." What's this?
1. If neutrons pile into nuclei very rapidly, before they can decay, the sequence of nuclei formed is very neutron rich.
2. Just as before, they tend to pile up at the neutron magic numbers of 50, 82, and 126.
3. However, if the neutron population suddenly goes away, then these nuclei decay back to make slightly lower elements in the sequence, around selenium, tellurium, and platinum.
4. These are the R-process peaks, and they lie just to the left of the S-process peaks.

D. Where do the S- and R-processes occur?
1. The S-process is thought to occur in very luminous red giant stars, where certain reactions maintain a low population of free neutrons.
2. The R-process is thought to occur during a core-collapse supernova, when a huge population of neutrons is rapidly generated.

V. A full theoretical treatment predicts 200 abundances that span a range of 1000 billion.

A. Amazingly, the measurements agree with almost all these to within a factor 2. This is one of 20th-century science's great triumphs.

B. But it's much more than that. It connects you to the Universe in a direct and powerful way.

Suggested Reading:

Ferris, *The Whole Shebang*, chap. 4.

Silk, *The Big Bang*, chap. 16.

Bartusiak, *Archives of the Universe*, chaps. 29, 46 (Payne and B^2FH papers).

Pagel, *Nucleosynthesis and Chemical Evolution of Galaxies*.

Questions to Consider:

1. How do we know what chemical elements are present, and in what relative amounts, in the stars and gas within the galaxy? Why might the element abundances we find on the Earth's crust be somewhat different from the cosmic average?

2. Sketch the famous cosmic abundance graph for all the elements, using an *exponential y*-axis spanning roughly 10^{12}. Identify the rare light elements, the alpha elements, the iron peak elements, and the 3 pairs of S- and R-process peaks. Compare this abundance graph with the periodic table, and check the elements you know to be common and rare.

3. There are two primary factors influencing this graph: the properties of atomic nuclei and the properties of stars. Considering both these, describe the origin of the various peaks you identified in the last question.

4. Look around you and notice the different elements in your surroundings, and in yourself. Try to imagine the birthplace for these different elements and their journey from there to here, just as you might imagine the birthplace for your car or your clothes in factories around the world.

Lecture Twenty-Four—Transcript
Understanding Element Abundances

In the last lecture, I gave an overview of how the atoms that are now inside you and me were originally made inside stars that lived and died several billion years ago. It's a wonderful story, and one of its qualities is that it connects us directly to the wider Universe. But you also may have noticed that I just told you the story; I didn't really include much evidence for it. How do we know that's what actually happened? Well, I hope by the end of this lecture you won't have any lingering doubts.

See, our approach is going to be to focus on the relative abundances of the various chemical elements. For example, the fact that iron is common and that gold is rare. So we're going to need to look at how element abundances are measured out in the galaxy. Then we'll look at the theory of nucleosynthesis to see what it predicts for these abundances. And then finally, it's the stunning match between theory and observation that makes this subject one of the great triumphs of 20th-century science.

So let's begin by seeing how astronomers figure out what stars are made of. It seems quite amazing that just by looking at a star you can actually tell what it's made of, but it turns out you can. Remember from Lecture Four, when a hot gas glows, its atoms give off light with a specific set of colors. Here are some examples. Each element glows with its own set of colors. It's a kind of "bar code" that acts like a spectral signature, a different bar code for each kind of atom. Now, in these examples, the colors are seen as bright or emission lines, but in other circumstances, the atoms can absorb these same colors, and the bar code appears as dark absorption lines against the bright, white light continuous spectrum.

Here, for example, is the full spectrum of the Sun, which shows the strongest lines. But a more detailed spectrum shows hundreds of lines. The bar codes from all the different elements are superposed. So obviously, by identifying the various bar codes in these spectra, you know what elements are in the star. Now, as it happens, measuring the relative amount of each element—the element's abundance—is much more difficult. It wasn't until 1925 that a young British-born Ph.D. student at Harvard Observatory, Cecilia Payne, finally developed a method to do this. And using this method, she was the first person to realize that the stars were primarily made of hydrogen and helium, and not iron as had been previously thought.

Now, in the 80 or so years since that time, the science of measuring cosmic abundances has become really quite mature, and we now have a very good idea of the composition of most of the objects in the Universe, whether they're like the stars here, or gas clouds like these. So what do we find for the cosmic abundances? You can actually boil the results down to three primary conclusions. First, almost always, one finds about three-quarters of hydrogen and one-quarter of helium, and a small amount of all the other elements, somewhere between 0% and 4%. Without doubt, the primary elements in the Universe are hydrogen and helium, while all the other elements are an afterthought.

Now, this result might puzzle you a bit because, obviously, the Earth certainly isn't made mostly of hydrogen and helium, but the Earth is an exception. Its gravity is too weak to hold onto these 2 light gases, so it lost them long ago, and it's made almost entirely from the other element category. The second result is that the overall fraction of heavy elements, that 0% to 4%, varies from object to object in a very interesting and systematic way that we'll look at in the next lecture. But the third result is that the relative abundances within the heavy elements are very similar in all objects, and this is going to be one of our key themes for this lecture. So let's look at these relative abundances now.

Remember, the most common way to see all the elements is arrayed in the famous periodic table, all 92 of them. But to astronomers and physicists, this graph of their relative abundance is much more interesting. The *x*-axis runs along the element series, from hydrogen to uranium, and the *y*-axis shows the relative amount, or abundance, of each element. Now, notice, the *y*-axis is exponential and spans a huge range. For example, for every 10 billion atoms of hydrogen, we have 1 million atoms of silicon, 100 atoms of zinc, and just 1 atom of lead. Its overall shape has a steady decline, with a number of peaks along the way. Hydrogen and helium are by far the most abundant elements. Then there's a precipitous factor of a billion drop to the light elements—lithium, beryllium, and boron—which are extremely rare. Then comes what is called the "alpha peak," with familiar elements—such as carbon, oxygen, neon, and magnesium—and then another dip before a major peak, the iron peak, in which iron and nickel really stand out. From then on, it's a gradual decline to extremely rare elements. But superimposed on this declining slope are 3 pairs of smaller peaks, which, for reasons we'll get to, I've labeled "R" and "S."

Overall, this should jibe with your own knowledge of the elements. You've heard of the common ones—oxygen, carbon, silicon, and iron—because they're common. While the rare ones, we've often never heard of—niobium, molybdenum, and dysprosium—and the few times you have heard of them—gold, silver, and platinum—is precisely because they're rare. I mean, that's why they're valuable. But remember what I mentioned a minute ago—this abundance curve is found wherever you look: stars, planets, interstellar gas. It's always the same. This is a wonderful example of our world mirroring the cosmic world. Gold isn't just rare here on Earth—it's rare everywhere. And carbon and oxygen aren't just common on Earth—they're common everywhere.

Now, having introduced the abundance curve, we're going to spend the rest of this lecture trying to understand its shape. The overall framework, of course, we met last lecture. Thermonuclear fusion reactions deep inside stars make these elements, but there's much more to the story than I gave you, and it's remarkable how all those peaks can be explained by the properties of atomic nuclei and star furnaces. So let's start with the first step along a long road to heavier elements, from hydrogen to helium. In the last lecture, I said that protons simply collide and stick, but in truth this is an exceedingly difficult reaction for protons to achieve because the reaction doesn't involve the strong nuclear "Velcro" force, but instead the weak force of radioactivity, and this makes the reaction extremely unlikely.

Here's the actual reaction: 2 protons react to make deuterium. It's a heavy version of hydrogen, with 1 neutron, and an antielectron and a neutrino are ejected. This reaction is so rare that in the core of the Sun, it takes the average proton a billion years to react, despite its colliding with another proton a billion times every second. So the Sun's furnace burns its hydrogen fuel incredibly slowly. A cubic foot of your digestion is 400 times more powerful than a cubic foot of the Sun's furnace. It's an extremely gentle furnace. The Sun is so powerful only because its furnace is so big. As you might imagine, it's because the hydrogen fuel is consumed so slowly that the Sun and stars have very long lives—in the case of the Sun, nearly 10 billion years. This slowness of star lives has a crucial cosmological significance. If stars evolved much faster, galaxies would have gone dark long ago, and there'd be no time for life to evolve.

Okay, back to our reactions. Once deuterium is formed, things proceed more rapidly. Deuterium absorbs another proton to make helium-3, a version of helium with only 1 neutron. Then 2 of these collide to make normal helium-4. Overall, these reactions have effectively combined 4 protons into a helium nucleus, and that's the nuclear ash from this first hydrogen fuel. So far, so good, but what about the next stage? How do we go beyond helium?

It turns out that this is also extremely difficult, for the simple reason that the two obvious next steps are blocked. Now, to help see this, I want to introduce an exceedingly useful kind of diagram that places all nuclei on an x, y plane. Up the y-axis, we have the sequence of elements, defined by the number of protons, and along the x-axis, we have the number of neutrons. So every nucleus can be placed on this graph. The unstable ones, I've colored purple. Now, you'll notice that each element has several possible nuclei. For example, lithium, with 3 protons, has 2 versions, or "isotopes," to use the proper word. Lithium-6 has 3 neutrons; and lithium-7, with 4 neutrons. Notice the number in the name refers to the total number of protons and neutrons.

Let's see what options we have for helium-4. Well, adding a proton would shift it up 1 place to lithium-5, but that nucleus doesn't exist, so that reaction isn't possible. What about colliding 2 helium-4 nuclei together? That would move 2 spaces up and 2 spaces to the right, landing on beryllium-8, but this is an exceedingly unstable nucleus. It lives for just 10^{-16} seconds before breaking apart back into 2 helium-4 nuclei, so that reaction is also blocked. Now, what actually happens is rather remarkable and was first figured out by Ed Salpeter and Fred Hoyle in the early 1950s. Nothing can happen until the helium core of a star crushes down to reach a density of many kilograms per cubic centimeter, and almost 100 million K. Only then is it possible for 3 helium nuclei to collide almost simultaneously to make carbon-12, in what's called the "triple alpha reaction" because another name for a helium nucleus is an "alpha particle."

In effect, the first 2 helium nuclei collide to make beryllium-8, but before it has a chance to break apart, a third one comes zooming in to make carbon. Well, we're making rather slow progress at the periodic table. We've got as far as element number 6, carbon, but things go a little faster from now on. In fact, historically, everything fell into place in 1957 with a landmark paper by this group of 4, Margaret and Geoff Burbidge, Willi Fowler, and Fred Hoyle, and their paper is now viewed as a classic, and is often referred to simply as "B^{2}FH," from their initials. In addition to hydrogen fusion and the triple alpha reactions, they spelled out a number of other nuclear pathways that generated, essentially, all the other elements up to and beyond uranium. So much of the story I'm about to tell you has its roots back in their paper.

Let's return to a larger version of our graph of possible nuclei. You can see down on the lower left, we've already gone from hydrogen to helium, via deuterium, at 10 million K. Then the triple alpha reaction takes us from helium to carbon-12 at 100 million K. And at that point, the reactions get a little easier. You see, the helium nucleus, the alpha particle, is such a robust little entity that it keeps getting added to nuclei to make a sequence called, not surprisingly, the "alpha elements." See here, we jump by 2 protons and 2 neutrons from carbon-12 to oxygen-16 to neon-20 to magnesium-24 to silicon-28 and to sulfur-32. All these elements are quite abundant.

Back to our abundance graph, let's zoom in on this alpha peak. The alpha elements are the red dots. You can see they have abundances typically between 10 and 100 times higher than the elements in between, things like fluorine, sodium, aluminum, phosphorus, chlorine, and potassium. Now, you might wonder how to get these elements in between. Well, at even higher temperatures, more like a billion K, then other reactions occur—for example, carbon-carbon or oxygen-oxygen fusion creates a much wider range of elements. Now, remember the reason these reactions need higher temperature—the nuclei have more protons, and so you need higher temperature to overcome their much stronger repulsion.

These conditions are getting pretty drastic, so let's see where all this is heading. Remember the all-important binding energy curve from the last lecture—it shows the glue energy stored in the nucleus. As long as a reaction moves downwards, then energy is released, and a star can still shine. But you can see the problem here. The fusion reactions yield less and less energy as they move down towards the bottom of the curve. Ultimately, when they get to iron, element number 26, that's it, game over—no more energy is available. Iron nuclei are the ultimate ash from fusion reactions. So without going into details, I think you can see how fusion reactions tend to generate large quantities of the elements around iron. And sure enough, this is exactly what's found.

Let's go back to our abundance graph. The iron peak stands out from elements 24 through 28: chromium, manganese, iron, cobalt, and nickel. They're all cosmically quite abundant, with iron significantly more than the others. Okay, this is great. I mean, we've managed to make elements up to the iron peak, but at this point, our path is blocked. Normal fusion can't generate heavier elements, and we need a radically different mechanism. This is where neutrons enter our story in a big way. I mean, in a sense, they come to our rescue. So let's see how they do this.

Here's yet another part of our neutron-proton graph. We're further along the sequence here, with iron-56 at the lower left. It has 26 protons and 30 neutrons. Iron-56 is called a "seed nucleus." Now, in certain stars, some reactions make free neutrons, and because they have no charge, they can sneak into any nucleus quite easily. So let's add a neutron to our iron-56 seed nucleus. We make iron-57. Add another, we make iron-58, and add 1 more to make iron-59. Now, iron-59 is unstable, and 1 of its neutrons changes into a proton by spitting out an electron. Bingo, we move up and to the left to make cobalt-59, a new element. The process continues. Adding a neutron to cobalt-59 makes cobalt-60, which in turn decays to nickel-60. So do you see what's going on? By steadily adding neutrons, we can sneak up the sequence, following this zigzag path, making unstable nuclei, which then decay by neutron conversion into protons.

Now, as we move up this path, a fascinating thing occurs. When we get to 50 neutrons, the elements begin to build up. Basically, they don't like to absorb another neutron, and when they do, they quickly decay back to the next element up, keeping the number of neutrons at 50. Why do nuclei like to have 50 neutrons, and not 49 or 51? Well, the answer is really quite delightful. You see, the protons and neutrons aren't just sort of glued together like the red and blue spheres in my diagrams—they're actually on orbits within the nucleus. And just like electrons orbiting within atoms, they orbit within shells. You may remember from high school chemistry that when electrons fill a shell in an atom, that atom is one of the noble gases, things like helium or neon or krypton, on the far right of the periodic table. These atoms are particularly stable because their electron shells are filled.

Well, exactly the same kind of pattern happens inside atomic nuclei. Protons and neutrons orbit within shells, and when a shell is filled, that nucleus is especially stable. The particular numbers when this happens are called "magic numbers," and the nuclei are called "magic nuclei." So now, let's step back and look at our entire neutron-proton diagram, from hydrogen all the way up and beyond uranium. In black, you can see all the stable nuclei. They curve over to the right because, remember, stable nuclei tend to have more neutrons than protons. Now, either side of the stable black nuclei are unstable nuclei. To the right are neutron-rich nuclei that decay by spitting out electrons, moving back to the stable nuclei.

In fact, you can understand why physicists call this the "nuclear valley of stability." The stable nuclei lie along the valley floor, and the walls of the valley rise on either side. And naturally, nuclei tend to roll down the walls, back to the valley floor. Now, here are the magic numbers. They're at 2, 8, 20, 28, 50, 82, and 126. Nuclei with this number of neutrons or protons are particularly stable and robust. They're the nuclear equivalent of the noble gas atoms. So now, let's look again at our neutron pathway. Up it comes, and where it meets these magic numbers, the elements build up—here, near zirconium, barium, and lead. But wait a minute; let's go back to our abundance diagram. That's exactly where we have peaks on the abundance plot. Here they are, at just the right location. You'll notice I've labelled them "S-process" to distinguish them from the other 3 peaks that are close by, called "R-process" peaks. Why are there 2 sets of peaks? The answer is it's all to do with at what frequency the neutrons are added. "S" stands for slow, and "R" stands for rapid.

See, here's our close-up of the S-process path. Here's the line, or valley, of stability. Just a little way from that line, the nuclei are only mildly unstable, and it takes days to years for them to decay. So as long as the neutrons only get absorbed infrequently, say, 1 every 1000 years, then each new nucleus has time to decay. That's what defines this particular path, the S-process path. Now, let's consider this region, further away from the valley of stability, up the valley walls a bit. The nuclei are now much more unstable. They take only fractions of a second to decay. So now imagine our seed nucleus was subject to a very rapid bombardment of neutrons, getting smacked many times per second. This time, the neutrons would pile in and push the iron into a very neutron-rich location, perhaps iron-66, before it would decay. So the path with rapid neutron addition would be more like this green line.

Let's back out to see the larger picture again. Here's the calculated path for an iron seed nucleus in a billion-degree neutron gas of roughly water density. You can see how the path stays far to the right of the stable nuclei, and the nucleus makes rapid progress, passing the magic numbers at 1/10, 1/2, and 2.0 seconds, and even reaching such heavy nuclei that fission breaks the nucleus apart and cycles it back round. Notice how just like the S-process pathway, the R-process accumulates many nuclei around the 3 magic neutron numbers, 50, 82, and 126.

Now, if the neutron flux shuts off, all those nuclei decay. They roll back down to the valley floor from the valley wall, neutrons converting into protons by spitting out electrons. They move diagonally to the left and up, so those regions of excess at the magic numbers end up here, as excesses near the elements selenium-80, tellurium-130, platinum-192, and even up at thorium and uranium. As you can already guess, these correspond perfectly to the second set of abundance peaks on the abundance graph. Here are the R-process peaks, offset slightly from the S-process peaks.

Now, so far, I've only referred to the S- and R-processes. Where might we find the right conditions for these mechanisms? Well, the slow path is thought to occur in very luminous red giant stars, where certain reactions make free neutrons. Over thousands of years, these neutrons get absorbed by heavier nuclei and populate the slow S-process pathway. In fact, we know this happens because the spectra of these stars show enhanced S-process elements, including the smoking gun, element number 43, technetium. Now, this element has no stable nuclei. It decays within a million years, so it must have been made recently, before getting dredged up onto the surface, where it leaves its mark in the spectrum.

The site of the R-process, the rapid addition of neutrons, is thought to be core-collapse supernovae. I don't have time to go into details, but as the core collapses, it forms a neutron star. So for a few seconds, there is a Herculean population of neutrons coming out of the core and into all the surrounding matter. It's a huge population of neutrons that drives the nuclei up the neutron-rich R-process pathway. A few seconds later, when the explosion is over, the nuclei then start decaying back to the line of stability.

Now, at this point, all I've done is to show you a few of the mechanisms that generate the various features on the abundance graph: the production of helium and carbon, the alpha and iron peaks, and the S- and R-processes. In truth, this subject is much more advanced than I've portrayed it: The lives of all the types of stars are calculated, including all the nuclear reactions that fuel their furnaces; the dredge-up and ejection of the elements that have folded in; and the slow, multigenerational buildup of elements in the galaxy is modeled; and of course, the observation of abundances in all kinds of locations gets ever more refined. The bottom line is that there is a stunningly good match between the calculated and the observed element abundances.

Let's not forget there are over 200 abundance values, 1 for each isotope, and these values range over a factor of 1000 billion. The calculations get essentially all of them correct to better than a factor of 2. This is amazingly good, and should convince you that we really do understand the origin of the elements. And from any perspective, this is one of the great achievements of 20th-century science. But, of course, it's more than a scientific success story—it's a fundamental part of the cosmological account of how we came to be here. Perhaps more than any other of the lectures in this course, this one forges the closest link between you and the Universe.

You now know enough to trace the story of most of the matter that's around you back to its origins. Within my hand, for example, the iron that makes the blood red—those atoms were most likely born in a white dwarf supernova. The carbon atoms that make up my fingernails were born in a Sun-type star. And the gold in my wedding ring—it sits right next to platinum on the R-process peak. These atoms have actually experienced the billion-degree chaos of a massive star's core collapse. They know what it is to be hurled at one-tenth light speed out into space. All of these atoms have witnessed the deep cold between the stars. They're all older than the Earth, and they'll all outlive the Sun.

If you ever feel weighed down by the banality of life, take a look at your surroundings. Take a look at yourself, and try to feel the cosmic in the local.

Math-Physics Appendix

Although the lectures in this course are primarily discursive, they also contain a few simple mathematical results. Almost always, a lecture format is too rapid to properly convey mathematical work, so this appendix allows you to review at your own pace a number of the more quantitative themes discussed in the lectures, hopefully deepening and extending your understanding. Much of this appendix is straightforward, but in parts the level is higher and is intended for folks who are still fluent with advanced high school level math and physics. You do *not*, however, need to know or follow *any* of this to appreciate the material discussed in the lectures.

The material is arranged according to the lecture in which it appears first.

Lecture Two: Denizens of the Universe

numbers and powers of 10: Here's a simple review of numbers and powers of 10, starting with a table of the most common prefixes. They are separated by a factor of $1000 = 10^3$, which is *large*. For example, $1000 compared to $1 is the same as a trillion dollars compared to a billion dollars, and a billion dollars compared to a million dollars.

Name	Number	Power of 10	Prefix	Example
trillion	1,000,000,000,000	10^{12}	tera-	TeV
billion	1,000,000,000	10^9	giga-	Gly
million	1,000,000	10^6	mega-	Mly
thousand	1000	10^3	kilo-	kg
	1	1		
thousandth	0.001	10^{-3}	milli-	mm
millionth	0.000001	10^{-6}	micro-	μm
billionth	0.000000001	10^{-9}	nano-	nm
trillionth	0.000000000001	10^{-12}	pico-	ps

For large numbers, the number of zeros is the power of 10 (e.g., $100,000 = 10^5$), while for small numbers it is the number of shifts of the decimal point (e.g., $0.00001 = 10^{-5}$). A negative power of 10 is also 1 over the positive power. For example: $10^{-5} = 1/10^5$.

Using powers of 10 in algebra is very easy: For multiplication, division, and exponentiation just add, subtract, and multiply the powers. For example, $10^4 \times 10^2 = 10^6$; $10^4 \div 10^2 = 10^2$; and $(10^4)^3 = 10^{12}$. Also, square roots are powers of 1/2, for example, $\sqrt{10^4} = (10^4)^{1/2} = 10^2$.

speed of light: $c \approx 3 \times 10^8$ m/s $\approx 3 \times 10^5$ km/s $\approx 10^{13}$ km/yr. These are all speeds in slightly different units. The last one tells us that a light-year is about 10 trillion km in length. More impressively, it is 7 times around the world in 1 second.

density: ρ = mass/volume. Units are kg/m^3 (or equivalent). The density of water is a standard: 1 g/cm^3 = 1000 kg/m^3 = 1 ton/m^3. Air has low density (~1 kg/m^3) while lead has high density ($\rho_{lead} = 11 \times \rho_{water}$). For cosmic densities, we often use units of hydrogen atom masses per cubic meter: m_H/m^3, where $m_H = 1.67 \times 10^{-27}$ kg.

exponential axes: Extremely useful in science, because you can fit very small and very large things on the same graph. With exponential axes, the same distance along the axis corresponds to the same *multiplicative* factor. Thus, we might have 10^2 to 10^3 to 10^4 to 10^5 all equally spaced.

Lecture Three: Overall Cosmic Properties

number counts: The lecture refers to the "minus 3/2 slope" test for cosmic evolution. More specifically, for a *uniform* distribution of galaxies, the number brighter than brightness b is $N_{>b} \propto b^{-3/2}$. As we include fainter galaxies, b *decreases*, so their number, $N_{>b}$, *increases*.

Why the $-3/2$ power? Well, the number you count depends on the *volume* of the survey, so $N \propto V \propto d^3$, where d is the depth of the survey. But fainter galaxies are further away: $b \propto 1/d^2$ (inverse square law, see Lecture Six), or equivalently: $d \propto 1/\sqrt{b} \propto b^{-1/2}$. So combining $N \propto d^3$ with $d \propto b^{-1/2}$, we have $N \propto (b^{-1/2})^3 \propto b^{-3/2}$, the "minus 3/2 power law." A graph with exponential axes for $N_{>b}$ and b gives a straight line with gradient $-3/2$. This reasoning also applies to a population of galaxies with a *range* of intrinsic luminosity, because it applies separately for galaxies of *each* luminosity.

The real data for radio galaxies rise *above* the expected line of slope $-3/2$ for small b, showing there are more faint galaxies, hence more galaxies at large distance, and therefore more galaxies *in the past*. Hence the population of radio galaxies is *changing*, and the Steady-State Theory cannot be correct.

Lecture Four: The Stuff of the Universe

atomic size: Most atoms are about 0.1 nm (10^{-10} meters) across. If these become marbles (1 cm = 10^{-2} m across), we multiply all sizes by 10^8, so a human cell (10^{-5} m) becomes $10^{-5} \times 10^8 = 10^3$ m = 1 km across, while a human (2 m) becomes 2×10^8 m = 200,000 km tall, and so on.

temperature scales: Kelvin = centigrade + 273.16; centigrade = (Fahrenheit − 32.0) × 5/9. So Kelvin and centigrade degrees have the same "size," and 0 K ($-273.16°C$) is the point at which all atomic and molecular motion stops.

temperature and atomic motion: Temperature measures the degree of motion in the atomic and molecular world. Specifically, it measures the *average* kinetic energy of the particles: $<KE> = \frac{3}{2}k_B T$, where k_B is Boltzmann's constant and T is the temperature in Kelvin. Particles of mass m at temperature T have a distribution of speeds, v, given by the Maxwell-Boltzmann curve: $N(v) \propto v^2 \exp(-mv^2/2k_B T)$.

light: An electromagnetic wave, with wavelength in the range 300 nm (violet) to 700 nm (red). The wavelength, frequency, and speed are related by $\lambda \times f = c$. One can also think of light as made of "particles" called photons, each with an energy $E = h \times f$, where h is Planck's constant.

thermal spectrum: The thermal motion of atoms (or other charged particles) creates photons with a spread of energies, giving the famous thermal (or Planck, or black-body) spectrum: $I(E) \propto E^3/[\exp(E/k_B T) - 1]$, where $I(E)$ is the brightness at photon energy E. Hotter objects glow with bluer light: $\lambda_{peak} \propto 1/T$ (called Wein's law), and they glow more brightly: $E_{total} \propto T^4$ (called the Stefan-Boltzmann law).

energy-mass equivalence: The most famous equation in all of physics expresses the deep equivalence of matter and energy: $E = mc^2$, where c is the speed of light. In one system of units, this becomes: E (in megatons) $= 0.97 \times m$ (in kg). You can think of matter as an exceedingly condensed and stable form of energy. (A megaton is a million tons of TNT explosive energy.)

Lecture Five: The Sweep of Cosmic History

scale factor: Written "S" in these lectures (also called the size, or separation factor), scale factor gives the separation of any 2 points (e.g., galaxies) relative to their separation today. S changes with time, with $S = 0$ at the Big Bang and $S = 1$ today. Lecture Twelve shows how the full expansion history, $S(t)$, depends on the Universe's density parameters: Ω_{rad}, Ω_{mat}, Ω_{vac}.

expansion and temperature: Because of red shift (see Lecture Seven), the thermal spectrum of photons shifts in such a way that its temperature follows: $T \propto 1/S$, where S is the cosmic scale factor. Because today $S = 1$ and $T_{CMB} = 2.725$ K, then we also have $T = 2.725/S$.

Lecture Six: Measuring Distances

slim triangle equation: For a slim triangle: distance $= (57.3 \times$ base)/(apex angle in degrees). This equation can be used either to calculate the distance to an object knowing the base and the angle, or the physical size of an object if the distance and angle are known.

inverse square law: At a distance d, an object of luminosity L has a brightness $b = L/4\pi d^2$. If L is known, and b is measured, then d can be calculated using this equation.

Lecture Seven: Expansion and Age

Doppler effect: When we see (or hear) an object that's moving away from us, its waves of light (or sound) are stretched by an amount $\Delta\lambda = \lambda_{obs} - \lambda_{emit}$. For light, this is a shift to the red; for sound, this is a drop in pitch. The amount of stretch is given by $\Delta\lambda/\lambda_{emit} = (\lambda_{obs} - \lambda_{emit})/\lambda_{emit} \approx v/c$, where c is the speed of light (or sound).

Hubble law: Nearby, a galaxy's Doppler velocity in km/s depends on its distance: $v = H_0 \times d$, where $H_0 \approx 22$ km/s/Mly is Hubble's constant, and d is the distance in Mly. At times other than today, $v = H \times d$, where H is the Hubble parameter ($H = H_0$ today). In reality, this law holds true to arbitrarily large distances and rates of separation, but we can only observe an approximation to the law because we're forced to witness objects in the past.

red shift: Light waves are also stretched by cosmic expansion, which astronomers usually parameterize by $z = \Delta\lambda/\lambda_{emit}$. If expansion is cast as a velocity, then for low red shift, $v \approx c \times z$, and $d \approx cz/H_0$.

redshift stretch factor: In cosmology, it is often better to cast red shift directly in terms of cosmic stretch, so for redshift stretch factor (*rsf*): $rsf = \lambda_{now}/\lambda_{emit} = d_{now}/d_{emit}$, where d_{now} and d_{emit} are the distances today and when the light set out. In terms of the scale factor, $S(t_{emit}) = d_{emit}/d_{now} = 1/rsf$. From their definitions, $rsf = 1 + z$. These lectures use *rsf* rather than z because of its slightly simpler connection to S.

Hubble time and true age: $t_H = 1/H$ is called the Hubble time and is a rough estimate of the age of the Universe ($t_{age} < t_H$ for deceleration, and $t_{age} > t_H$ for acceleration). Today, $t_H \approx 13.5$ Byr (using $H_0 = 22$ km/s/Mly). The true cosmic age, t_{age}, is given by $S(t_{age}) = 1$ and is 13.7 Byr for the concordance parameters (the competing influence of the deceleration and acceleration phases makes t_{age} quite close to t_H).

Hubble distance: The distance light travels in the Hubble time is called the Hubble distance, or Hubble radius: $r_H = c \times t_H = c/H$. It is a rough estimate of the size of the visible Universe.

Lecture Eight: Distances, Appearances, and Horizons

three distances: In an expanding Universe, we can define 3 distances to an object—d_{now}, d_{dmit}, and d_{LT}—for the current distance, the distance when the light set out, and the distance light itself has traveled. These can be calculated from the red shift using the expansion history $S(t)$. First, find t_{emit} using $S(t_{emit}) = 1/rsf$, so that $d_{LT} = c \times (t_{now} - t_{emit})$; then, using calculus: $d_{now} = c \times \int 1/S(t)\, dt$ where the integral goes from t_{emit} to t_{now}; finally: $d_{emit} = d_{now} \times S(t_{emit}) = d_{now}/rsf$.

horizon distance: The current distance to an object whose light set out at the Big Bang: $d_{horizon} = d_{now}$ for $t_{emit} = 0$, so $d_{horizon} = c \times \int 1/S(t) \, dt$ with the integral from 0 to t_{now} ($\approx$ 46 Bly for the concordance parameters). You can think of it as the current distance to the galaxies that formed from the gas that we now see as the microwave background.

event duration: An event occurring in a distant galaxy appears to take longer: duration observed = $rsf \times$ true duration. This is one explanation for cosmic red shift: The observed duration between successive wave crests of light is longer than the true duration at the object.

object brightness: The apparent brightness of a distant object is given by $b = 1/4\pi d_{now}^2 \times 1/rsf^2$. The first term accounts for how light spreads over a large sphere of radius d_{now}; the second term accounts for red shift weakening each photon and slowing its rate of arrival.

object size: The apparent angular size of a distant object is given by $\theta° = 57.3 \times size/d_{emit} = 57.3 \times rsf \times size/d_{now}$. This is just the slim triangle equation (Lecture Six) using d_{emit} for the distance.

Lecture Nine: Dark Matter and Dark Energy—96%!

circular orbital motion: Newton's second law of motion is $F = m \times a$ (force = mass $\times$ acceleration). In circular orbital motion, the force is provided by gravity: $F = GMm/R^2$, while the resulting centripetal acceleration is $a = V^2/R$. Inserting these into $F = ma$ gives $GMm/R^2 = m \times V^2/R$ or $M = RV^2/G$. Measuring orbital speed, V, in km/s and orbit radius, R, in light-years gives the mass, M, in solar masses responsible for making the orbit: $M = 72 \times R_{ly} \times V_{km/s}^2$ solar masses.

rotation curves: For circular orbital motion, $V(R)$ is called the rotation curve, and it reveals the distribution of mass within a system. For the solar system (all mass at the center): $V \propto 1/\sqrt{R}$. For a galaxy with $V = $ constant, mass is spread out, with $M \propto R$.

Lecture Ten: Cosmic Geometry—Triangles in the Sky

The next three lectures are mathematically the most challenging by far. Don't worry if you can't follow the math. But if you can, it's really quite rewarding.

Euclidean geometry: Here are the standard rules of Euclidean geometry. Circle circumference: $C = 2\pi r$; Sphere's area: $A_{sphere} = 4\pi R^2$; Sphere's volume: $V_{sphere} = \frac{4}{3}\pi R^3$; sum of interior triangle angles $= 180°$.

circles on a sphere: On a sphere, a circle's circumference is smaller than the Euclidean result: $C = 2\pi R \sin(r/R)$, where R is the radius of the sphere, and r is the arc from the circle center to the circle (in this formula, r/R is in radians, not degrees). This is also the result for a circle within a non-Euclidean space with curvature radius R.

critical density: The critical density that makes the Universe's geometry flat is $\rho_{crit} = 3H^2/8\pi G$. Its value today (using $H_0 = 22$ km/s/Mly) is 9.7×10^{-27} kg/m^3 = 5.8 m_H/m^3.

omegas: Cosmologists express cosmic densities relative to the critical density; these are called "omegas." For example, the average density of atomic matter is about 0.25 m_H/m^3, so we have $\Omega_{at} \approx 0.25/5.8 = 0.04$. What affects cosmic geometry is Ω_{tot}, the sum of all the omegas. Only if $\Omega_{tot} = 1$ is the geometry Euclidean.

curvature radius: Although not given in the lectures, there is a simple formula for the Universe's curvature radius: $\mathcal{R}^2 = r_H^2/|\Omega_{tot} - 1|$. For $\Omega_{tot} = 1$, $\mathcal{R} = \infty$ (flat geometry), while for $\Omega_{tot} = 1.01$, $\mathcal{R} = 10 \times r_H \approx 135$ Bly. From current observations, $\mathcal{R} > 135$ Bly.

Lecture Eleven: Cosmic Expansion—Keeping Track of Energy

kinetic, gravitational, and total energies: Kinetic energy, $KE = \frac{1}{2}mV^2$ is the energy associated with a mass m moving at speed V; Gravitational energy, $GE = -GMm/R$, is the energy needed to pull two masses (M and m) from separation R to infinity (also called "binding energy"); TE is total energy. In a simple gravitating system, $TE = KE + GE$. Because GE is always negative and KE is always positive, TE can be either positive (the system is unbound) or negative (the system is bound).

escape velocity: In a marginally bound system, $TE = 0$, so $\frac{1}{2}mV^2 = GMm/R$, giving a critical escape velocity $V_{esc} = \sqrt{(2GM/R)}$. For a sphere with uniform density ρ expanding with a Hubble law, we may substitute $V = H \times R$ and $M = \frac{4}{3}\pi R^3 \rho$ and square both sides:

$$H^2R^2 = 8\pi GR^3\rho/3R.$$

Canceling R and rearranging, we get an expression for the critical density $\rho = \rho_{crit}$:

$$\rho_{crit} = 3H^2/8\pi G.$$

Friedmann equation: In the lecture, we skipped the following derivation of the Friedmann equation for the expanding rock sphere. The energy of the outer rock is conserved during its trajectory:

$$\frac{1}{2}mV^2 - GMm/R = \frac{1}{2}mV_0^2 - GM_0m/R_0 = TE,$$

where subscript "0" signifies the starting (current) time. Cancel the m's, multiply through by 2, and substitute $M_0 = \frac{4}{3}\pi R_0^3 \rho_0$ and $M = \frac{4}{3}\pi R^3 \rho$:

$$V^2 - \frac{8}{3}\pi GR^2\rho = V_0^2 - \frac{8}{3}\pi GR_0^2\rho_0.$$

Divide through by V_0^2:

$$(V/V_0)^2 - 8\pi GR^2\rho/3V_0^2 = 1 - 8\pi GR_0^2\rho_0/3V_0^2.$$

Substitute the Hubble law, $V_0 = H_0R_0$, and the scale factor, $S = R/R_0$:

$$(V/V_0)^2 - 8\pi GS^2\rho/3H_0^2 = 1 - 8\pi G\rho_0/3H_0^2.$$

Substitute $8\pi G/3H_0^2 = 1/\rho_{0,crit}$ to get our final Friedmann equation:

$$(V/V_0)^2 - (\rho/\rho_{0,crit}) \times S^2 = 1 - \rho_0/\rho_{0,crit}.$$

After specifying $\rho(S)$, a simple algebraic solution to this equation gives $V(S)$, the expansion velocity as a function of size of the sphere.

expansion history, $S(t)$ and $V(t)$: In the lecture, we skipped how to go from $V(S)$ to the expansion history, $S(t)$ and $V(t)$. Here's how that's done. Start from the normal energy equation:

$$\tfrac{1}{2}mV^2 - GMm/R = TE.$$

Divide through by m, and use $R = R_0 \times S$ to rearrange:

$$V = [2GM/SR_0 + TE/m]^{1/2}.$$

From calculus, $V = dR/dt = R_0 dS/dt$. Substitute this, bring R_0 across, and define the right side to be $f(S)$:

$$dS/dt = [2GM/SR_0^3 + TE/mR_0^2]^{1/2} \equiv f(S).$$

So we now have a simple differential equation:

$$dS/f(S) = dt.$$

Integrate this from $S = 0$ at $t = 0$ (the Big Bang) forward in time:

$$\int dS/f(S) = t.$$

In practice, the integral can be tricky, but ultimately one can rearrange the result to get S as a function of time, $S(t)$. Differentiating this, we get: $dS/dt = V(t)/R_0$, and using $V_0 = H_0 R_0$ and $t_H = 1/H_0$, we get the velocity history:

$$V(t)/V_0 = t_H dS(t)/dt.$$

An example follows.

$S(t)$ and $V(t)$ for the critical case: A critically expanding, pure matter Universe has $TE = 0$, so $f(S) = \sqrt{(2GM/SR_0^3)} = (2GM/SR_0^3)^{1/2}$.

Plug this into $dS/f(S) = dt$ and rearrange to get:

$$S^{1/2}dS = (2GM/R_0^3)^{1/2}dt.$$

Integrate from $S = 0$ at $t = 0$:

$$\tfrac{2}{3}S^{3/2} = (2GM/R_0^3)^{1/2}t.$$

Substitute $M = \tfrac{4}{3}\pi R_0^3 \rho_{0,\text{crit}}$, then cancel R_0^3 and rearrange:

$$S^{3/2} = (\tfrac{8}{3}\pi G\rho_{0,\text{crit}})^{1/2}\, t.$$

©2008 The Teaching Company.

Substitute $\rho_{0,\text{crit}} = 3H_0^2/8\pi G$ and $t_H = 1/H_0$, and bring the $^2/_3$ across:

$$S^{3/2} = {}^3/_2 H_0 t = {}^3/_2\, t/t_H.$$

Finally, invert this to get the expansion history, $S(t)$:

$$S(t) = ({}^3/_2\, t/t_H)^{2/3}.$$

Notice that overall $S \propto t^{2/3}$, and today, for $S = 1$, $t_{\text{now}} = {}^2/_3\, t_H$.

The velocity history is $V(t)/V_0 = t_H \mathrm{d}S/\mathrm{d}t = t_H({}^3/_2)^{2/3}(1/t_H)^{2/3} \times {}^2/_3\, t^{-1/3}$:

$$V(t)/V_0 = {}^2/_3({}^3/_2)^{2/3} \times (t/t_H)^{-1/3} = 0.87 \times (t/t_H)^{-1/3}.$$

So overall, we have deceleration with $V \propto t^{-1/3}$.

Friedmann's original equation: Friedmann derived in 1922 the following equation for an isotropic, homogeneous Universe obeying Einstein's General Relativity:

$$(\mathrm{d}S/\mathrm{d}t)^2 - {}^8/_3\, \pi G\rho \times S^2 = -kc^2/\mathcal{R}_0^2,$$

where k is -1, 0, or $+1$ for open/flat/closed geometries, and $\mathcal{R}_0$ is the current spatial curvature radius. In the lecture, we skipped the algebra that took us from this equation to our familiar Newtonian version, so let's do that here. First, combine $V = H \times R$, $V = \mathrm{d}R/\mathrm{d}t$, and $S = R/R_0$ to get $\mathrm{d}S/\mathrm{d}t = S \times H$. Substitute that, and $c^2/\mathcal{R}_0^2 = H_0^2 \times |\Omega_{\text{tot}} - 1|$ and combine k with $|\Omega_{\text{tot}} - 1|$:

$$S^2 H^2 - {}^8/_3\, \pi G\rho \times S^2 = -H_0^2(\Omega_{\text{tot}} - 1).$$

Now divide by H_0^2 and substitute ${}^8/_3\, \pi G/H_0^2 = 1/\rho_{0,\text{crit}}$:

$$S^2 H^2/H_0^2 - (\rho/\rho_{0,\text{crit}}) \times S^2 = 1 - (\rho_0/\rho_{0,\text{crit}}).$$

Now use the Hubble law: $V = H \times R$, and $V_0 = H_0 \times R_0$:

$$(S^2 V^2/R^2)/(V_0^2/R_0^2) - (\rho/\rho_{0,\text{crit}}) \times S^2 = 1 - (\rho_0/\rho_{0,\text{crit}}).$$

Finally, substitute $R/R_0 = S$ and simplify to get the Newtonian form:

$$(V/V_0)^2 - (\rho/\rho_{0,\text{crit}}) \times S^2 = 1 - (\rho_0/\rho_{0,\text{crit}}).$$

Recall that General Relativity duplicates a Newtonian analysis in this case because of Birkhoff's Theorem: An isotropic distribution of mass outside a sphere has no effect on the matter inside the sphere.

Lecture Twelve: Cosmic Acceleration—Falling Outward

density evolution: As the Universe expands, the average density of matter, radiation, and dark energy are: $\rho_{mat} \propto 1/S^2$, $\rho_{rad} \propto 1/S^4$, $\rho_{de} =$ constant. Matter is simply diluted as the volume increases; radiation is also redshifted, while dark energy (vacuum) density doesn't change. The total density is $\rho_{tot} = \rho_{mat} + \rho_{rad} + \rho_{de}$.

full Friedmann equation: If we insert ρ_{tot} into the Friedmann equation from Lecture Eleven and recall that $\rho_{mat} = \rho_{0,mat}/S^3$ etc., and $\rho_{0,mat}/\rho_{0,crit} = \Omega_{mat}$ etc., we get:

$$(V/V_0)^2 - (\Omega_{mat}/S^3 + \Omega_{rad}/S^4 + \Omega_{de}) \times S^2 = 1 - \Omega_{tot}$$

or

$$\boxed{(V/V_0)^2 - (\Omega_{mat}/S + \Omega_{rad}/S^2 + \Omega_{de}S^2) = 1 - \Omega_{tot}.}$$

times of equality: The scale factor, S_{eq}, at which the matter and radiation densities are equal occurs when $\Omega_{mat}/S_{eq}^3 = \Omega_{rad}/S_{eq}^4$, giving:

$$\boxed{S_{eq} = \Omega_{rad}/\Omega_{mat} \approx (0.000084/0.27) = 0.00031 \text{ (or } t = 57{,}000 \text{ yr).}}$$

Similarly, the scale factor at which the matter and dark energy densities are equal occurs when $\Omega_{mat}/S_{eq}^3 = \Omega_{de}$:

$$\boxed{S_{eq} = (\Omega_{mat}/\Omega_{de})^{1/3} \approx (0.27/0.73)^{1/3} \approx 0.72 \text{ (or } t = 9.4 \text{ Byr).}}$$

dark energy—falling outward: One can gain a Newtonian understanding of dark energy's *accelerating* expansion by carefully considering what "falling" really means. When something falls, it moves to a position of *lower gravitational energy*. For example, 2 masses "fall together" because their gravitational energy, $E_{grav} = -GMm/R$, *decreases* (gets more negative) when R *decreases*. A stone falls to the ground because E_{grav} gets more negative.

Similar reasoning applies to a sphere of matter, with gravitational energy $E_{grav} \approx -GM^2/R$ (actually, it's $^3/_5$ times this, but let's ignore the factor of $^3/_5$ here). Because M is fixed, then $E_{grav} \propto -1/R$, so if R *decreases*, E_{grav} gets more negative, so a sphere of matter *falls inward*.

However, a sphere of dark energy behaves differently because its mass is *not* fixed when it expands. Because dark energy has constant density, then the sphere's mass $M \propto R^3$, so $E_{grav} \propto -R^3 R^3/R \propto -R^5$, so if R *increases*, E_{grav} gets more negative, so a sphere of dark energy *falls outward*. Basically, outward is "downhill."

The situation is a bit more complicated because the gravitational energy liberated by falling must also make the new dark energy. In Lecture Thirty-Three, we learn this only happens if the sphere is larger than the Hubble sphere.

alternate explanation: There is an alternate explanation for dark energy's accelerating expansion, which you may have already come across. It describes dark energy as having *negative pressure*, which causes gravity to be repulsive. The explanation may also refer to Friedmann's *second* equation, in which acceleration is proportional to −(mass-energy density + 3 × pressure), so that a negative pressure can reverse the sign and turn gravity from attractive to repulsive.

While both explanations are ultimately equivalent, the second is, to my mind, extremely difficult to grasp intuitively. The fact that pressure makes gravity is already difficult to intuit, but the fact that *negative* pressure (tension) pushes outward is almost beyond intuition.

The difference between the two explanations boils down to whether one chooses to specify the changing energy density directly using things like $\rho_{rad} \propto S^{-4}$, or $\rho_v \propto S^0$ (= constant) or indirectly using pressure; and on whether one chooses to think about what adds to energy (Friedmann's first equation) or what adds to acceleration (Friedmann's second equation). My own view is that the first approach is by far the more intuitively accessible.

S(t) in the radiation era: In the lecture, we learned that expansion in the radiation era follows $S(t) \propto \sqrt{t} \propto t^{1/2}$. Let's derive that by first inserting into the Friedmann equation $\Omega_{mat} = \Omega_{vac} = 0$ and $\Omega_{rad} = 1$:

$$\boxed{(V/V_0)^2 - 1/S^2 = 0.}$$

Take $1/S^2$ to the right and take the square root of both sides:

$$V/V_0 = 1/S.$$

Now use $V = dR/dt = R_0 dS/dt$ and $V_0 = H_0 R_0 = R_0/t_H$:

$$R_0 dS/dt = R_0/(t_H S).$$

Cancel R_0 and rearrange:

$$S dS = dt/t_H.$$

Now integrate from $S = 0$ at $t = 0$ (the Big Bang), forward in time:

$$\tfrac{1}{2} S^2 = t/t_H.$$

Finally, rearrange to get $S(t)$:

$$\boxed{S(t) = 2^{1/2}(t/t_H)^{1/2} \propto t^{1/2}.}$$

Notice that today ($S = 1$), $t_{now} = 2^{-1/2} \times t_H = 0.71 \times t_H$, which is less than t_H, as we expect for decelerating expansion.

S(t) in the dark energy era: In the same way, one can derive dark energy's exponential expansion, $S(t) \propto 2^t$, by inserting $\Omega_{mat} = \Omega_{rad} = 0$ and $\Omega_{vac} = 1$ into the Friedmann equation, giving:

$$\boxed{(V/V_0)^2 - S^2 = 0.}$$

Following the previous analysis, this leads to:

$$dS/S = dt/t_H.$$

Which is integrated from $S = 1$ at $t = t_{now}$ to give:

$$\log_e S = (t - t_{now})/t_H$$

or

$$\boxed{S(t) = \exp[(t - t_{now})/t_H].}$$

This can be expressed as $S \propto 2^y$, where $y = 1.4 \times t/t_H$, giving a doubling time of ~1.4× the Hubble time (technically, t_H is the "e-folding time").

time of coasting: The transition between deceleration and acceleration occurs when the velocity (squared) curve is at its minimum. This occurs at the *peak* of the gravitational curve. To find this, we look for where its slope is zero (the curve is horizontal). The gravitational curve is given by $\Omega_{mat}/S + \Omega_{rad}/S^2 + \Omega_{de}S^2$. Ignore Ω_{rad}/S^2, differentiate to find the slope, and set this to zero:

$$-\Omega_{mat}/S^2 + 2\Omega_{de}S = 0$$

$$2\Omega_{de}S = \Omega_{mat}/S^2$$

$$(S_{coast})^3 = \Omega_{mat}/2\Omega_{de}$$

$$S_{coast} = (\Omega_{mat}/2\Omega_{de})^{1/3} \approx (0.27/1.46)^{1/3} \approx 0.57 \text{ (or } t \approx 7.0 \text{ Byr)}.$$

By looking at the *slope* of the gravitational energy, we're effectively looking at the *force* experienced by the sphere. Coasting occurs when the inward force of the matter equals the outward force of dark energy (slopes are equal and opposite, summing to horizontal). Interestingly, kilogram for kilogram, dark energy has *twice* the force of matter, which is why the time of equal density occurs *after* the time of equal forces (i.e., coasting).

Lecture Thirteen: The Cosmic Microwave Background

the CMB dipole: Over cosmic history, the Milky Way galaxy has picked up a speed of about 600 km/s as it falls toward nearby galaxy clusters. This results in a Doppler pattern on the microwave background called a "dipole." The emission is slightly blueshifted in our direction of motion, slightly redshifted in the opposite direction, with a smooth change in between. Because Doppler shifting a thermal spectrum is equivalent to changing its temperature, the dipole pattern is often shown as a temperature map, with temperature change $\Delta T/T = v/c = 0.002$ for the maximum at 600 km/s.

Lecture Fourteen: Conditions during the First Million Years

tracking change: One can follow conditions across the first million years using the scale factor: $S \approx 1.2 \times 10^{-6} \times \sqrt{t_{yr}}$ in the radiation era, and $S \approx 1.5 \times 10^{-7} \times t_{yr}^{2/3}$ in the matter era. Roughly, at $t = 10^2$, 10^3, 10^4, 10^5, and 10^6 years, we have $S = 1.2 \times 10^{-5}$, 3.8×10^{-5}, 1.2×10^{-4}, 4.4×10^{-4}, and 1.5×10^{-3}.

density: Because cosmic volume $\propto S^3$, then total matter density = $\rho_{0,mat} \approx 2.6 \times 10^{-27}/S^3$ kg/m^3; particle (proton/electron) number density $n_e \approx 0.35/S^3$ per m^3; photon number density $n_{ph} \approx 410/S^3$ per cm^3.

temperature: Via red shift, the temperature changes: $T = 2.7/S$ Kelvin, passing ~3000 K near 400,000 years.

sky color and intensity: The wavelength (color) of the peak in the thermal spectrum is: $\lambda_{pk} = 2.9 \times 10^6/T \approx 10^6 \times S$ nm (recall: blue to red is 300 nm to 700 nm). The total sky intensity: $I_{sky} = 2.2 \times 10^{-7}T^4$ Watt/m^2 = $1.6 \times 10^{-10}T^4$ noon Suns. So at 100,000, 400,000, and 800,000 years, that's 280,000, 12,000, and 3000 noon Suns.

fogginess: Visibility distance is $1/n\sigma$, where n is the particle number density, and σ is the scattering cross section (e.g., πr^2 for a water droplet in a fog). In the early Universe, free electrons act like water droplets, with $\sigma = 6.6 \times 10^{-29}$ m^2, giving a visibility distance: $1/[0.35/S^3 \times 6.6 \times 10^{-29}]$ meters $\approx 5 \times 10^{12}\, S^3$ ly, or about 500 ly at 100,000 years. This suddenly increases near 400,000 years when the free electrons disappear to make atoms.

Lecture Fifteen: Primordial Sound—Big Bang Acoustics

gas pressure: $P \approx nk_BT \propto n \times T$, where n is the particle number density, T is the temperature in Kelvin, and k_B is Boltzmann's constant. In our air, $n \approx 10^{19}/$cm^3 and $T \approx 300$ K, while in the foggy gas at 100,000 years, photons are most numerous with $n \approx 3 \times 10^{12}/$cm^3 and $T \approx 6000$ K, so the pressure is $P \approx 10^{-6}$ atm. In this gas, pressure (sound) waves can form and move.

primordial sound: Arises when the primordial gas falls into and bounces out of regions (clumps) of slightly higher dark matter density. Since regions of all sizes exist, they act like an ensemble of roughly spherical organ pipes, from small (high pitch) to large (low pitch). Near the time of the CMB, a typical sound wave is ~20,000 ly across moving at ~$0.5c$ (see below), so it has a period of ~40,000 years and frequency ~10^{-12} Hz, requiring an upshift of ~50 octaves ($2^{50} \approx 10^{15}$) to enter the human audible range (~10^3 Hz).

drop in pitch: At any given time, only regions smaller than the sound horizon (the distance sound can travel since the Big Bang) support sound waves. As time passes and the sound horizon grows, ever-larger "organ pipes" are added to the register, and so the pitch drops. The rate at which the pitch drops depends on the speed at which the sound horizon grows, which in turn depends on the speed of sound.

sound speed: Roughly the average particle velocity. In air, $v_s \approx 330$ m/s, while for a pure photon gas it is $v_s \approx c/\sqrt{3} \approx 0.6c$. More exactly, $v_s^2 = c^2/(3 + 9\rho_{at}/4\rho_{ph})$, so as the photon to atomic density ratio decreases, the sound speed also decreases, so the expansion of the sound horizon slows down. Hence, the primordial sound pitch drops rapidly at first, then slows down. Ultimately, when the fog clears at 400,000 years, the sound speed plummets (to ~10 km/s), the sound horizon freezes, the pitch stabilizes, and true oscillating sound waves cease.

sound loudness: Decibels—dB $= 20\log_{10}(\Delta P/\Delta P_0)$—measure the ratio of the pressure amplitude of the sound wave, ΔP, to a reference pressure amplitude, ΔP_0. In air, $\Delta P_0 = 2 \times 10^{-10}$ atm, so waves of amplitude 2×10^{-10}, 10^{-9}, 10^{-6}, 10^{-3} atm have loudness 0, 15, 75, 135 decibels (0 dB is the quietest audible sound: a mosquito at 3 meters). In Big Bang acoustics, we use the same relative amplitude scale, so pressure waves of amplitude 10^{-9}, 10^{-6}, 10^{-3} of the ambient pressure have loudness 15, 75, 135 dB. The relation between pressure variations in the gas and temperature variations seen on the CMB is simple: $\Delta P/P = 4 \times \Delta T/T$, so temperature variations of $10^{-5} - 10^{-4}$ reveal a sound loudness of 105–125 decibels: in the rock-concert range.

sound spectra: The CMB patchiness can be described by a spectrum of waves, $C(\lambda)$, where $\lambda \approx 180/\theta°$ is *angular* frequency (waves per degree). Although $C(\lambda)$ contains many nonacoustic effects, one can calculate the *pure* sound spectrum, $P(k)$, where $k = 2\pi/\lambda$ is a *spatial* frequency (waves per Mly). Although $P(k)$ technically describes the density variations, because temperature is uniform (and $P \propto n \times T$), it also describes the pressure (sound) variations. There are actually *three* versions of $P(k)$, one each for atomic matter, dark matter, and photons, with $P_{at}(k)$ describing sound.

Lecture Seventeen: Primordial Roughness—Seeding Structure

roughness: Can be quantified as variations in density, $\Delta\rho/\rho$ from place to place. Using Fourier methods, one can reexpress $\Delta\rho/\rho$ as the sum of many wavelike components, whose strengths are given by a spectrum, $P(k)$, and whose offsets are given by phases $\theta(k)$, where $k = 2\pi/\lambda$ is spatial frequency. The cosmological situation is special because the phases are random and we can ignore them (hence they're not mentioned in the lecture). So $P(k)$ gives a complete description of cosmic roughness.

growth of roughness: Density variations can grow as the Universe ages in two different ways. Large-scale (super-horizon) regions grow by differential expansion, with $\Delta\rho/\rho \propto S^2 \propto t$ in the radiation era, and $\Delta\rho/\rho \propto S \propto t^{2/3}$ in the matter era; while smaller-scale (sub-horizon) regions grow by internal gravitational clumping, with $\Delta\rho/\rho \approx$ constant in the radiation era, and $\Delta\rho/\rho \propto S \propto t^{2/3}$ in the matter era. All roughness growth is suppressed in the dark energy era. In this context, the horizon size is $r_H = ct_H = c/H$, the Hubble sphere ($r_H = 2c\,t_{age}$ and $^3/_2\,c\,t_{age}$ in the radiation and matter eras).

initial and final roughness spectrum: The initial roughness spectrum injected by inflation is $P(k) \propto k$. As long as $\Delta\rho/\rho$ is small (say, < 0.1) then each k component grows independently of all other k's, so one can easily track how $P(k)$ evolves with time, as we did in the lecture. Today's roughness spectrum is as follows: unchanged, $P(k) \propto k$, on very large scales; a gentle peak for scales ~ 200 Mly; and falling, $P(k) \propto 1/k^2 - 1/k^3$, on smaller scales.

two forms for roughness, $P(k)$ and $\Delta(k)$: In this course, I have used exclusively $P(k)$ to describe cosmic roughness. It gives the amplitude (squared) of the "sine waves" that add to make the density distribution. In some contexts, cosmologists prefer to use $\Delta^2(k) \approx k^3 P(k)$, which has a different shape (no peak, just rising with k). $\Delta(k)$ describes roughness differently: It's the spread (standard deviation) of the average density inside a box of size $d \approx 1/k$ placed randomly at many locations. Be careful when looking at graphs labeled "power spectra" to check whether $P(k)$ or $\Delta(k)$ is plotted.

Lecture Nineteen: Infant Galaxies

number of visible galaxies: About 10,000 faint galaxies are found in the Hubble Ultra Deep Field. How does this translate to the *total* number visible all over the sky? The HUDF is a 6-arc-minute square. Now, on the Earth's surface, 1 arc minute is (by definition) 1 nautical mile (nmi). So the number of HUDF fields that cover the sky is the same as the number of 6×6 nautical mile squares that cover the Earth. Because the Earth's radius is 3963 miles = 3443 nmi, its surface area is $4\pi R^2 \approx 1.50 \times 10^8$ nmi^2. So $1.50 \times 10^8/36 \approx 4.1 \times 10^6$ HUDFs cover the sky, with $4.1 \times 10^6 \times 10^4 \approx 4 \times 10^{10}$ potentially visible galaxies. Hence the rough statement of "100 billion" (10^{11}) galaxies in the visible Universe.

Lyman break redshifts: The spectra of extremely distant (young) galaxies often include a sharp drop in the ultraviolet, $\lambda_{LB} \approx 100$ nm, called the "Lyman break." At high red shift, this break can appear in the optical and near infrared. For example, with $\lambda_{obs} = 400, 600, 800$, and 1000 nm, we have redshift stretch factors: $rsf = \lambda_{obs}/\lambda_{LB} = 4, 6, 8$, and 10 and corresponding ages: 2.2, 1.2, 0.77, and 0.55 Byr after the Big Bang.

Lecture Twenty: From Child to Maturity—Galaxy Evolution

star-star collisions: These are very rare, even when 2 galaxies collide. Why is this? A typical stellar density in a galaxy is $\sim 1/ly^3$. Now imagine a tunnel through a galaxy, with length $\sim 10^4$ ly and cross-section 1 ly^2 so it contains $\sim 10^4$ stars. When 2 galaxies collide, you can imagine *each* star passing through a tunnel of this kind, with a probability of striking one of the stars in the tunnel of $A_{star}/A_{tunnel} \times N_{stars} \approx (10^6)^2$ km$^2/(10^{13})^2$ km$^2 \times 10^4 = 10^{-10}$ (where a typical star diameter is $\sim 10^6$ km). With $\sim 10^{11}$ stars in a galaxy, maybe 10 will collide. So almost all stars are "safe" during the collision.

Lecture Twenty-One: Giant Black Holes—
Construction and Carnage

black hole size: Called the "Schwarzschild radius" and depends only on the mass of the black hole: $R_s = 2GM_{bh}/c^2$. In Newtonian terms, it's the radius at which the escape velocity: $V_{esc} = \sqrt{(2GM/R_s)} = c$, the speed of light. In units of the Sun's mass: $R_s = 3M_{bh}/M_{Sun}$ kilometers, so for the Sun, $R_s = 3$ km; for the Earth, $R_s = 3M_{earth}/M_{Sun} \approx 10^{-5}$ km = 1 cm; for a 10^8 M_{Sun} quasar, $R_s = 3 \times 10^8$ km = 2 AU (Earth's orbit diameter).

black hole energy release: Dropping mass m down to the Schwarzschild radius releases (positive) energy: $E \approx GM_{bh}m/R_s = GM_{bh}m/(2GM_{bh}/c^2) = \frac{1}{2}mc^2$, which is 50% of the mass-energy. In practice, the mass carries kinetic energy into the black hole, and this reduces the efficiency to $\sim 0.1mc^2$. If a black hole is fed at a rate of 1 M_{Sun}/yr, this releases a power of $0.1 \times (2 \times 10^{30}$ kg$) \times (3 \times 10^8$ m/s$)^2/(3.1 \times 10^7$ s/yr$) \approx 5 \times 10^{38}$ Watt $\approx 1.5 \times 10^{12}$ $L_{Sun} \approx 100$ L_{gal}, which is a typical quasar luminosity.

Lecture Twenty-Three: Atom Factories—Stellar Interiors

atomic nuclei: They are 1 – few femtometers across (fm = 10^{-15} meter), 10^5 times smaller than an atom; similar to a fly in a football field. Because nuclei contain almost all the atom's mass, the density of nuclear matter is about $(10^5)^3 = 10^{15}$ times *higher* than the density of atomic matter, or about 10^{15} gm/cm^3, compared to 1 gm/cm^3. The mass of all humans is $\approx 6 \times 10^9 \times 80$ kg $\approx 5 \times 10^{14}$ gm, which is roughly 1 cm^3 of nuclear matter. Hence the statement: Remove the space within atoms, and the human race fits inside a sugar cube.

nuclear and chemical energy: One can trace the million-fold increase in efficiency of nuclear fuel compared to chemical fuel to the fact that atoms are 100,000 times larger than nuclei. This means the electrostatic energy of the nuclear protons is $\sim 10^5$ times larger than the atomic electrons. A further factor of 10 arises because an outer electron that's involved in a chemical reaction doesn't feel the full nuclear charge, being shielded by the inner electrons. Quantitatively, rearranging chemical electrons releases ~ 1eV/atom, while rearranging nuclear protons and neutrons releases ~ 1MeV/atom. Notice we need only consider electrostatic energy because in the nucleus it's comparable to the nuclear (strong) energy.

red giant atmospheres: When a star like the Sun ends its life, it swells up huge and loses its atmosphere. Part of the reason is that the gravitational force holding the atmosphere down is so much weaker in the larger star. For example, using $F_{grav} \propto M/R^2$, the Sun's current surface gravity is 30 times Earth's gravity, but when it swells to fill the Earth's orbit, its gravity is over 10,000 times weaker—only just enough to prevent the atmosphere from simply evaporating off into space.

white dwarf supernova bomb: Here's a rough estimate of the energy released by a white dwarf supernova. It involves the fusion of 1.4 M_{Sun} of carbon and oxygen. If the reactions make iron, then our nuclear binding energy curve tells us that each proton mass liberates about 1 MeV of energy. The total is therefore: $E \approx 1.4 \times M_{Sun}/m_p \times 1$ MeV $\approx 3 \times 10^{44}$ joules $\approx 6 \times 10^{28}$ megatons—a pretty big explosion!

core collapse supernova bomb: Here's a rough estimate of the energy released by a core collapse supernova. The energy is primarily gravitational: a core of ~1.4 M_{Sun} collapses to make a neutron star with radius $R_{NS} \approx 20$ km. The gravitational binding energy of such a sphere is $E \approx {}^3/_5 GM^2/R \approx 1.6 \times 10^{46}$ joules $\approx 4 \times 10^{30}$ megatons, of which 99% emerges as neutrinos, leaving ~10^{44} joules of explosive (kinetic) energy.

Lecture Twenty-Four: Understanding Element Abundances

slow hydrogen fusion: In the lecture, the slowness of the proton + proton reaction in the Sun was explained as due to its dependence on the weak nuclear force. There is another, equally important, reason: at only 15 million K, protons never collide and touch! Let's find their closest approach by equating their average kinetic energy to their electrostatic repulsion energy at closest approach: $\frac{1}{2} m_p <v^2> = {}^3/_2 k_B T = Q^2/4\pi\varepsilon_o R_{min}$, where Q is the proton charge. Solving for $R_{min} \approx 2Q^2/12\pi\varepsilon_o k_B T \approx 750$ femtometers. Even for head-on collisions with each proton moving at 3 times the average speed, we still only get $R_{min} \approx 40$ fm, or ~50 proton radii (a similar relative separation between the Moon and Earth). Why do they ever react? Classically, at 15 million K they would never collide and touch, but in quantum mechanics they can "tunnel" through the gap with very low, but finite, probability.

Lecture Twenty-Five: Light Elements—Made in the Big Bang

conditions near 1 minute: The Friedmann equation tells us that in the radiation era $S(t) \approx 2.1 \times 10^{-10} t_{\text{sec}}^{1/2}$. So, at 20 seconds, when $S \approx 10^{-9}$ we have: $T = 2.72/S \approx 2.7$ billion K; $\rho_{\text{at}} = \rho_{0,\text{at}}/S^3 \approx 430$ gm/m^3; $\rho_{\text{rad}} = \rho_{0,\text{rad}}/S^4 \approx 815$ tons/m^3 (photons + neutrinos); sky color $\lambda_{\text{peak}} = 2.9 \times 10^6/T \approx 0.003$ nm (γ-rays); sky intensity $I_{\text{sky}} = 1.6 \times 10^{-10} T^4 \approx 1.6 \times 10^{26}$ noon Suns; and the composition was $1/7/7/10^{10}$, for neutrons/protons/electrons/photons. The minute-old Universe was, with some important differences mentioned in the lecture, rather similar to the center of a present-day star.

deuterium bottleneck: Nuclear fusion reactions only start when deuterium (H-2, pn) can survive. Because its binding energy is 2.2 MeV, one naively expects it to survive below $T = 25$ billion K (because at that temperature, $<E> \approx k_{\text{B}}T \approx 2.2$ MeV), which occurs at ¼ second. However, because there are so many photons compared to protons, even at much lower temperatures there are still many photons with energy above 2.2 MeV capable of destroying deuterium. Only below 1 billion K (~3 minutes) does the fraction of photons with energy above 2.2 MeV become sufficiently low (<10^{-9}) for deuterium to finally survive. Hence nucleosynthesis is "put on hold" until ~3 minutes.

Lecture Twenty-Six: Putting It Together— The Concordance Model

dark energy's "w" parameter: Defines how the density of dark energy changes with expansion, S. Specifically: $\rho_{\text{de}} = \rho_{0,\text{de}}/S^{3(1+w)}$, so for $w < -1$, ρ_{de} decreases with expansion; for $w = -1$, ρ_{de} is constant; for $w > -1$, ρ_{de} increases with expansion. Current measurements suggest $w \approx -1$.

concordance parameters: Here's the table of cosmological parameters given in the lecture, with approximate uncertainties. Not all these values and uncertainties are independent. The values pertain to 2006, but as the data are improved and expanded, the values may change slightly.

Parameter	Sym	Value	Error (%)
expansion rate	H_0	22 km/s/Mly	5
dark energy	Ω_{de}	0.73	5
dark matter	Ω_{dm}	0.23	5
atomic matter	Ω_{at}	0.044	5
photons	Ω_{ph}	5.0×10^{-5}	0.1
neutrinos	Ω_{nu}	3.4×10^{-5}	10
geometry ($\Sigma \, \Omega_i$)	Ω_{tot}	1.00 (Euclidean)	2
photon/proton ratio	η	1.6×10^9	5
roughness index	n	0.96	2
roughness amplitude	A	0.81	3
dark energy "w"	w	-1.02	10
equality ($\rho_r = \rho_m$)	t_{eq}	57,000 yr	5
recombination	t_{cmb}	380,000 yr	2
reionization (CMB)	t_{reion}	400 Myr	15
age	t_{age}	13.7 Byr	1

concordance $S(t)$: Using the concordance omegas above, here are some approximations to $S(t)$ that are correct to <10% within the specified ranges. Radiation era: $S(t) \approx 1.16 \times 10^{-6} t_{yr}^{1/2} \approx 2.1 \times 10^{-10} t_{sec}^{1/2}$ ($t < 15$ kyr); matter era: $S(t) \approx 1.5 \times 10^{-7} t_{yr}^{2/3}$ (1 Myr $< t < 10$ Byr); dark energy era: $S(t) \approx \exp[(t - t_{now})/15.9]$ with t in Byr ($t > 10$ Byr).

Lecture Twenty-Seven: Physics at Ultrahigh Temperatures

energy units: A unit of energy commonly used in subatomic physics and cosmology is the electron volt, eV, or its larger cousins, the MeV or GeV (10^6 or 10^9 eV). The eV is tiny by human standards, only 1.6×10^{-19} joules, which is the physicist's more usual unit of energy (roughly equal to 1 kg dropped through 4 inches). For hot gases (e.g., the early Universe), the temperature defines the average particle or photon energy: $E_{MeV} \approx 1.29 \times 10^{-10} T_K$ (using $<E> = \frac{3}{2} k_B T$), hence either temperature or energy can be used to refer to the conditions of the gas.

threshold temperatures: Each particle has a specific mass and mass-energy. For example protons have mass: $m_p = 1.67 \times 10^{-27}$ kg, and mass-energy, $m_p c^2 = 938$ MeV. If 2 particles collide with combined energy greater than $2 \times m_p c^2$, then a proton and antiproton can be created out of the collision energy. Thus, for each particle, there is a threshold temperature, $k_B T_{th} \approx mc^2$ (k_B is Boltzmann's constant), above which these particles and their antiparticles can be created naturally from thermal particle collisions in the gas. Some threshold temperatures are: electron ($\sim 4 \times 10^9$ K or 0.5 MeV); muons ($\sim 8 \times 10^{11}$ K or 100 MeV); pions ($\sim 10^{12}$ K or 140 MeV); protons ($\sim 10^{13}$ K or 940 MeV).

standard model particles: The 3 tables below give the matter (fermion) and force (boson) particles of the standard model, while the third table gives some composite particles. For each particle, an antiparticle also exists (with a few particles being their own antiparticle). In the fermion table, the 2 quantities given are the particle's mass and electric charge.

Fermions: spin ½, matter particles								
	1st family			2nd family			3rd family	
Leptons	electron e	0.51 MeV	-1	muon μ^-	106 MeV	-1	tau τ	1.8 GeV	-1
	e–neutrino ν_e	≈ 0	0	μ–neutrino ν_μ	≈ 0	0	τ–neutrino ν_τ	≈ 0	0
Quarks	up u	≈ 3 MeV	$+\frac{2}{3}$	charm c	≈ 1.3 GeV	$+\frac{2}{3}$	top t	171 GeV	$+\frac{2}{3}$
	down d	≈ 6 MeV	$-\frac{1}{3}$	strange s	≈ 100 MeV	$-\frac{1}{3}$	bottom b	≈ 4.2 GeV	$-\frac{1}{3}$

Lecture Twenty-Eight: Back to a Microsecond— The Particle Cascade

neutron-to-proton ratio: When neutrons and protons keep interconverting due to neutrinos, their ratio is given by the famous Boltzmann equilibrium formula $N(n)/N(p) = \exp(-\Delta E/k_B T)$, where ΔE is the difference in energy between the neutron and proton: $\Delta E = \Delta m \times c^2 = 939.5 - 938.3 = 1.2$ MeV. So at temperatures below 1.2 MeV $\approx 10^{10}$ K, the neutron fraction begins to plummet. This would occur after a few seconds, but at ~1 second, neutrinos decouple and the reactions freeze at $N(n)/N(p) \approx 1/6$. If neutrinos decoupled at 100 seconds, the ratio would have been 10^{-8}—essentially no neutrons at all!

neutron decay: Free neutrons decay ($n \rightarrow p + e + \bar{v}_e$) with a half-life of ~15 minutes. In the 3-minute delay during the deuterium bottleneck, about 13% of the neutrons decay: $(\frac{1}{2})^{3/15} \approx 0.87$. This changes their initial ratio of 1/6 at ~1 second to 1/7 at the time of nucleosynthesis.

Lecture Twenty-Nine: Back to the GUT—
Matter and Forces Emerge

electrostatics unmasked: It is difficult to grasp the enormous power of Nature's forces. The strong and weak nuclear forces only act across minute distances, and the electrostatic force is hugely muted by thorough intermixing of positive and negative charges. To unmask this, let's follow the example given in the lecture. Separate the electrons and protons from 2 sand grains 1 meter apart. Each is about 1 mm across and weighs ~3 mg, with ~10^{21} protons and electrons. In electrostatics, 10^{21} electrons carry a charge of ~300 Coulombs, so using "Coulomb's law": $F = Q^2/4\pi\varepsilon_o d^2$, with $Q = 300$, $d = 1$, and $1/4\pi\varepsilon_o = 10^9$, we have a force of ~$10^{14}$ newtons, or about ~10^{13} lb, or ~10 billion tons, or ~8 million space shuttles pulling in opposite directions. On macroscopic scales, with full charge separation, the electrostatic force is stunningly powerful.

the particle desert: In the standard model, not much happens between electroweak unification at $t = 10^{-14}$ seconds and Grand Unification at $t = 10^{-36}$ seconds. This is an utterly enormous gulf of conditions: a factor of 10^{22} in time corresponds to 10^{11} in scale factor (recall $S \propto t^{1/2}$ in the radiation era), and therefore 10^{11} in temperature and 10^{44} in energy density. This is an equivalent exponential range to the following: 10 μs to the age of the Universe, a million times colder than liquid helium to a billion-degree stellar core, interstellar vacuum to an atomic nucleus. If no changes occur over that kind of range of conditions, then this period is indeed an impressive "desert."

Lecture Thirty: Puzzling Problems Remain

fine-tuned escape speed: Here's the math that shows why the expansion speed must be closer and closer to the escape speed at earlier and earlier times. From the lecture: Launch a rock of from radius R_{ej} (ej = eject) at speed V_{ej} so that it eventually turns around at R_{turn}. The Newtonian equation for kinetic, gravitational, and total energy of the rock is:

$$\tfrac{1}{2}V_{ej}^2 - GM/R_{ej} = TE = -GM/R_{turn},$$

where the right side is the total energy when the rock is stationary at turnaround, and we've already divided through by the rock's mass, m. Now here's the same equation for the marginal escape condition:

$$\tfrac{1}{2}V_{esc}^2 - GM/R_{ej} = TE = 0, \text{ so that } \tfrac{1}{2}V_{esc}^2 = GM/R_{ej}.$$

Divide this into the first equation:

$$(V_{ej}/V_{esc})^2 - 1 = -(R_{ej}/R_{turn}).$$

Now take square roots, rearrange and write V_{ej} as $V_{esc} - \Delta V$:

$$(V_{esc} - \Delta V)/V_{esc} = 1 - (\Delta V/V_{esc}) = [1 - (R_{ej}/R_{turn})]^{1/2}.$$

Now assume R_{ej}/R_{turn} is small and invoke the approximation:

$$[1 - R_{ej}/R_{turn}]^{1/2} \approx 1 - \tfrac{1}{2}R_{ej}/R_{turn},$$

and we arrive at:

$$\Delta V/V_{esc} \approx \tfrac{1}{2} R_{ej}/R_{turn}.$$

So the fraction below escape speed is roughly half the ratio of the launch radius to the final turnaround radius. Hence the statement in the lecture: "if we're within 1% of escape speed now, we were within 0.1% of escape speed when the Universe was 10 times smaller." This analysis applies during the matter era, but the condition in the radiation era is even stricter.

evolving omega: In the lecture we used an equation for Ω to see how deceleration drove Ω away from 1, while acceleration drove it toward 1. Here's how we got that equation. First, notice that: $\Omega = \rho/\rho_{crit}$ and $\rho_{crit} = 3H^2/8\pi G$, so $\rho_{crit}/\rho_{0,crit} = H^2/H_0^2$ where, as usual, subscript "0" means today's values. Now use the Hubble law: $V_0 = H_0 R_0$ and $V = H R = HSR_0$ (using $S = R/R_0$) to arrive at $V/V_0 = S \times H/H_0$. So combining these we find:

$$S^2/\rho_{0,crit} = (V/V_0)^2/\rho_{crit}.$$

Now take the Friedmann equation: $(V/V_0)^2 - (\rho/\rho_{0,crit}) \times S^2 = 1 - \Omega_0$ and substitute in for $S^2/\rho_{0,crit}$ to get:

$$(V/V_0)^2 - (\rho/\rho_{crit}) \times (V/V_0)^2 = 1 - \Omega_0.$$

Now substitute for $\rho/\rho_{crit} = \Omega$ and divide through by $(V/V_0)^2$ to get:

$$1 - \Omega = (1 - \Omega_0)/(V/V_0)^2,$$

which gives our final equation for Ω:

$$\Omega = 1 - [(1 - \Omega_0)/(V/V_0)^2].$$

It's this equation that clearly shows how Ω diverges from 1 for deceleration (V/V_0 decreasing) and converges on 1 for acceleration (V/V_0 increasing). Hence, any mechanism that causes acceleration (e.g., inflation) will launch an $\Omega = 1$ Universe at the escape velocity.

Lecture Thirty-One: Inflation Provides the Solution

gravitational timescale: $\approx 1/\sqrt{(G\rho)}$ gives the approximate orbital, collapse, or expansion timescale for any system whose motion is dominated by gravity. It is a key relation in astrophysics and was introduced in the lecture without justification. Here's a simple derivation, using circular orbital motion from Lecture Nine. Equating the centripetal and gravitational forces, we have:

$$mV^2/R = GMm/R^2.$$

We also have the orbit period (circumference/speed):

$$P = 2\pi R/V$$

and the average density within the orbit (mass/volume):

$$\rho = M/({}^{4}\!/_{3}\pi R^{3}).$$

Substituting for P and ρ in the first equation (first canceling m), we get:

$$P^2 = 3\pi/G\rho$$

or

$$\boxed{P \approx 1/\sqrt{(G\rho)},}$$

where "$\approx$" means "forget about factors of 3 and π." In this example, the orbit period, P, is the gravitational timescale of the system.

This also works for cosmic expansion, where $\rho \approx \rho_{crit} \approx 3H^2/8\pi G$, so $1/\sqrt{(G\rho)} = 1/\sqrt{(3H^2/8\pi)} \approx 1/H = t_H \approx t_{age}$, the age of the Universe. For vacuum-dominated "outfall," $1/\sqrt{(G\rho)}$ gives the approximate doubling timescale. Recall from Lecture Twelve: $S(t) \propto \exp[(t - t_{now})/t_H]$, so S grows by a factor $e = 2.718$ every Hubble time: $t_H \approx 1/\sqrt{(G\rho)}$.

exponential expansion: In the lecture, we only discussed the minimum amount of inflation that solves the flatness and horizon problems. For example: if inflation occurs from 10^{-36} to 10^{-34} seconds, giving ~100 doublings, the increase in size is $\sim 2^{100} \approx 10^{30}$. But what if inflation ended at 10^{-30} seconds, so there were 1 million doublings? The increase in size would be $\sim 2^{\text{million}} \approx 10^{300,000}$. This would also be (within factors ~10^{30}) the relative size of the true Universe to our visible Universe. To get an idea of this, the relative size of the visible Universe to the size of a proton is 10^{10} ly $\times 10^{13}$ m/ly $\times 10^{15}$ fm/m $= 10^{38}$, which is *nothing* compared to $10^{300,000}$. Do you see why a significant period of inflation can generate a truly ("exponentially") huge Universe?

Lecture Thirty-Two: The Quantum Origin of All Structure

quantum uncertainty: Despite its classical appearance, physical reality in fact behaves in a quantum mechanical fashion. One of the features of this behavior is that certain pairs of properties cannot simultaneously take on precise values. Famous examples are position and momentum, and energy and time. The "fuzziness" of these quantities can be expressed using Heisenberg's uncertainty principle: $\Delta x \Delta p \approx h$; $\Delta E \Delta t \approx h$, where h is Planck's constant. This effect is important during inflation, because the energy density of the inflaton field isn't well defined. When the region expands through the Hubble sphere, this quantum fuzziness gets frozen as a real variation, and this is how cosmic roughness—from which galaxies will ultimately form—is generated.

Hubble radius during inflation: From the Hubble law: $V = H \times R$, so the Hubble radius (the distance at which $V = c$) is: $R_H = c/H$. Now, for a critical Universe: $H^2 = {}^8/_3 \pi G \rho$, so if ρ decreases, H decreases, and R_H increases. But during inflation, ρ is constant, so H is constant, so R_H is constant. For example, if inflation occurs at 10^{-36} seconds, then $t_H \approx 10^{-36}$ seconds and R_H is fixed at $\sim 10^{-36}$ light-seconds ($\sim 10^{-13}$ fm): *much* smaller than a proton.

the initial power spectrum: Imagine roughness being injected everywhere on scales of R_H, the Hubble radius. Because R_H is constant during inflation, the amplitude of the injected roughness remains constant. However, once outside the Hubble sphere, the accelerating expansion tends to *reduce* the density contrast by driving the local Ω to 1. The longer the time since exiting the Hubble sphere, the larger the region becomes, and the smaller the density fluctuation. So, when inflation ends, we have a roughness spectrum $P(k) \propto k$, with smaller $\Delta\rho/\rho$ on larger scales. This kind of *density* roughness has an interesting property: The variation in *gravitational potential energy*, $\Delta\phi/\phi$, is the *same* for all scales, and because ϕ defines the warping of space-time, then inflation generates a space-time landscape with equal "wrinkliness" on all scales. For this reason, the spectrum is often referred to as "scale-invariant."

random phases: One of the deep qualities of quantum fluctuations is their inherent randomness. During inflation, this emerges as randomness in the *location* of the "ups" and "downs" in the density field (technically, the phase of the Fourier waves is random). Because the *growth* of structures doesn't alter their location (except near turnaround and collapse), then this quantum quality of randomness propagates through to the present time and is visible in the weblike distribution of galaxies.

Lecture Thirty-Three: Inflation's Stunning Creativity

all is nothing: Here's a Newtonian analysis to show that the total energy in the visible Universe is zero. Its radius is $R_{VU} \approx R_H = c/H_0$; and its density is $\rho = \rho_{crit} = 3H^2/8\pi G$. So the total mass is (to within factors ~2):

$$M_{VU} = {}^4\!/_3 \pi R_{VU}{}^3 \times \rho \approx c^3/H_0 G \approx 10^{53} \text{ kg}.$$

The gravitational energy is:

$$E_{VU} \approx -GM_{VU}{}^2/R_{VU} \approx -G(c^6/H_0{}^2 G^2)/(c/H_0)$$
$$\approx -(c^3/H_0 G) \times c^2 \approx -10^{53} \text{ kg} \times c^2.$$

It seems that the positive mass in the Universe is (at least approximately) balanced by an equal negative gravitational energy. In General Relativity, the Newtonian gravitational energy is replaced by curved space-time, and the condition of zero total energy reemerges as the condition of zero curvature (i.e., Euclidean, or flat, geometry). Thus the measured flatness of the Universe is equivalent to it having zero net energy, which is what you expect for the energy-conserving mechanism of inflation acting on an initial tiny seed.

critical seed size: A sphere of vacuum energy will only fall outward if the gravitational energy released is great enough to make the new vacuum of the larger sphere. The condition is therefore:

$$GM_{sph}M_{shell}/R_{sph} > M_{shell} \times c^2.$$

Cancel M_{shell}, substitute $M_{sph} = \frac{4}{3}\pi R_{sph}^3 \rho_v$, and $\rho_v = 3H^2/8\pi G$, and simplify to get $\frac{1}{2}R_{sph}^2 H^2 > c^2$. Now, $1/H = t_H \approx t_{dbl}$: The Hubble time is roughly the expansion doubling time. So, ignoring factors of ~2, we have:

$$R_{sph} > R_H \approx c \times t_{dbl}.$$

For the threshold mass, we have $M_{sph} > R_{sph}c^2/G$, or:

$$M_{sph} > c^3/G \times t_{dbl} \approx 10^{35} \times t_{tbl} \text{ kg,}$$

where t_{dbl} is in seconds. Hence, for inflation near the GUT transition, with $t_{tbl} \approx 10^{-36}$ seconds, we need a seed vacuum region bigger than ~10^{-13} femtometers with mass greater than ~0.1 kg (roughly, a small apple).

Lecture Thirty-Four: Fine Tuning and Anthropic Arguments

Mandelbrot's equation: In the lecture, we compared the equation for a circle, $x^2 + y^2 = 1$, with a simple-looking equation that yields the famously complex Mandelbrot diagram. Here's the equation:

$$Z_{n+1} = Z_n^2 + C.$$

Now, the Z's and C are all "complex numbers," meaning they're of the form $a + ib$ where a and b are normal numbers while $i = \sqrt{-1}$. In the graph, the x-axis is for the "a" part and the y-axis is for the "b" part. The procedure takes every point C on the plane and generates a sequence of Z's of increasing n, starting with $Z_0 = 0$. Whether or not the sequence diverges ($|Z|$ goes to infinity) defines the boundary of the Mandelbrot diagram, and the speed of divergence defines the color. The boundary has an extremely complex form, which never simplifies, no matter how small a region is studied. In the current context, this is used to illustrate that simple algorithms can yield complex results. Perhaps, by analogy, a simple (yet to be discovered) symmetry principle in physics can yield our amazingly complex Universe.

Lecture Thirty-Five: What's Next for Cosmology?

comparing telescopes: In the lecture, we compared a number of different telescopes. In the hunt for yet-fainter galaxies, telescope aperture is a key figure of merit. For example, the mirrors in JWST and HST have diameters of 6.5 and 2.4 meters, so JWST represents an improvement by a factor $(6.5/2.4)^2 = 7.3$. Notice the ratio of diameters is *squared* to get the ratio of collecting areas. Let's repeat this to compare JWST with your eye, with pupil diameter ~5mm (in the dark): $(6.5/0.005)^2 = 1.7$ million. When you include the ratio of the sensitivity of the JWST camera to your retina, and the possibility of long exposure times, then the telescope as a whole is $\sim 10^{12}$ times more sensitive than your eye. With such amazingly enhanced vision, it perhaps isn't so surprising that we can now see objects in the remote Universe.

observing reionization: A major goal is to observe the 21-cm radio emission from neutral atomic hydrogen in the young Universe. This gas formed at recombination and was destroyed at reionization: ~200–800 Myr; or $rsf \approx 20 - 8$. The 21-cm emission arrives with wavelength between $20 \times 21 \approx 420$ cm and $8 \times 21 \approx 170$ cm, or frequencies ($f = c/\lambda$) between 71 MHz and 176 MHz. These are challenging frequencies to work at (mainly because of foreground emission), but progress is good and detection of primordial hydrogen hopefully isn't too far off.

Timeline

The following timeline provides a rough guide to the sequence of most events and processes discussed in these lectures. Although the primary designation is by time, equivalent properties are also given, such as scale factor (S), temperature (T, of the photons), average energy per particle (E), density (ρ, total unless otherwise indicated), and pressure (P). Some events may not occur (indicated by "?"), or their times are not well-known or not well-defined (indicated by "~"). Note: t = time, sec = second, m = meter, ly = light-year, Mly = million light-years, Bly = billion light-years, kyr = thousand years, Myr = million years, Byr = billion years, atm = atmosphere, $\propto$ = proportional to.

Time	Events/Epoch
10^{-43} sec	**Planck era**: A period of profound uncertainty, when space-time (gravity) takes on quantum form. The concept of smooth time may not apply, rendering the time assignment, relative to a hypothetical $t = 0$, as somewhat arbitrary. Compactification may occur near this time, defining the laws of physics.
10^{-36} sec	**GUT transition**: $S \sim 10^{-28}$, $T \sim 10^{28}$ K, and $E \sim 10^{16}$ GeV. Strong force may separate from electroweak force.
? 10^{-36} sec	**Inflation begins**: Possible time of inflation. A dense vacuum somehow forms and drives rapid exponential expansion ($S \propto 2^t$) for at least 50–100 doubling times, but possibly much longer. If inflation occurs at this time, the density is $\sim 10^{73}$ tons/cm^3, the doubling time is $\sim 10^{-36}$ sec, and the Hubble radius is $\sim 10^{-28}$ m. Quantum fluctuations

become real fluctuations with $P(k) \propto k$; these provide the seeds of cosmic roughness.

? 10^{-34} sec....................................**Inflation ends/reheating**: When inflation ends, the vacuum energy transforms to real energy (an event called reheating), leaving a critically expanding, Euclidean (flat) region containing all particles at a density $\sim 10^{73}$ tons/cm^3. The particles are relativistic, so the expansion changes from accelerating $\propto 2^t$ to decelerating $\propto \sqrt{t}$. The Standard Hot Big Bang has now been launched.

10^{-34}–10^{-5} sec..............................**Quark era/quark-gluon plasma**: From inflation to quark confinement, all particles are present. If only standard model particles exist, then quarks dominate in a quark-gluon plasma, and this interval is called the "particle desert."

? 10^{-17}–10^{-15} sec..........................**Supersymmetry breaks?** Many theories suggest the early Universe obeys Supersymmetry, which is broken around $E \sim 10$–100 TeV. Heavy-partner particles to the standard model are made at this time.

? $\sim 10^{-15}$ sec...................................**Lightest supersymmetric (dark matter) particle?** Theory suggests the lightest supersymmetric particle is stable with a mass ~ 10 TeV. This would be a WIMP (weakly interacting massive particle) and might be the (cold) dark matter particle.

$\sim 10^{-10}$ sec ....**Electroweak transition**: $S \sim 10^{-15}$, $T \sim 10^{15}$ K, and $E \sim 100$ GeV. The electroweak symmetry is broken; the weak and electromagnetic forces become distinct; the W and Z particles acquire mass, as do all particles, via the Higgs mechanism.

$\sim 10^{-5}$ sec ....**Quark confinement**: $S \sim 10^{13}$, $T \sim 10^{12}$ K, $E \sim 250$ MeV, and $\rho \sim$ nuclear density. The temperature and density are sufficiently low for quarks to bind to form hadrons. This marks the end of the quark era, and the end of the (possibly liquid) quark-gluon plasma.

$10^{-5} - 10^{-4}$ sec ....**Hadron era**: As quarks are confined, hadrons are formed, including protons, neutrons, pions, and their antiparticles.

$\sim 10^{-4}$ sec ....**Hadron annihilation**: A brief period of proton-antiproton and neutron-antineutron annihilation. Only one in a billion protons and neutrons survive; all the rest annihilate. The origin of the matter-antimatter asymmetry is not yet known, but probably occurred sometime earlier.

$10^{-4} - 10$ sec ....**Lepton era**: Following hadron annihilation, lightweight particles (such as electrons and positrons) dominate the energy density.

1 sec	**Neutrino decoupling**: $S \sim 10^{-10}$, $T \sim 10^{10}$ K, $E \sim 1$ MeV, and $\rho \sim 1$ ton/cm^3. The neutrinos no longer interact with the matter because the density and reaction probability have dropped too low. Neutrinos now freely stream, making the Cosmic Neutrino Background (CNB). The neutrino decoupling freezes the neutron/proton ratio near 1/6, which sets the stage for nucleosynthesis.
10 sec	**Electron annihilation**: $S \sim 3 \times 10^{-9}$, $T \sim 3 \times 10^{-9}$ K, and $\rho \sim 5$ kg/cm^3. Electrons and positrons annihilate, leaving behind the one-in-a-billion resulting from the matter-antimatter asymmetry. The total number of remaining electrons and protons is now equal.
10 sec–57 kyr	**Radiation era**: After electron annihilation, the dominant component is light (photons): soft gamma rays at 10 sec, X-rays from 100 seconds to 600 years, and UV at 60 kyr. During this time, $S \propto \sqrt{t}$ and roughness on sub-horizon scales is inhibited from growing.
1–5 min	**Nucleosynthesis**: $S \sim 10^{-9}$, $T \sim 10^9$ K, and $\rho_{at} \sim 10$ gm/m^3. As soon as deuterium can survive, nuclear reactions make helium-4. The initial neutron/proton ratio is $\sim 1/7$, giving 25% helium by mass, along with $\sim 10^{-5}$ deuterium and helium-3. These abundances help measure cosmic proton + neutron density (giving $\Omega_{at} \sim 0.04$).

57 kyr ....**Matter radiation equality**:
$S \sim 3 \times 10^{-4}$, $T \sim 9000$ K,
$\rho \sim 10^5$ m$_H$/cm^3, and $P \sim 10^{-5}$ atm.
The radiation (photon + neutrino)
density and matter (dark + atomic)
density are equal near 57,000 years.
Expansion changes from $S \propto t^{1/2}$ to
$S \propto t^{2/3}$, allowing dark matter
structure to grow on sub-horizon
scales ($\delta\rho/\rho \propto S$). Atomic matter is
held by light's pressure and
undergoes acoustic oscillations,
bouncing in and out of the dark
matter clumps.

380 kyr ....**Recombination**: $S \sim 10^{-3}$,
$T \sim 3000$ K, and $P \sim 10^{-7}$ atm.
Spanning an interval of ~80 kyr, the
free electrons and nuclei pair up to
form atoms. As a result, the fog
lifts; the gas pressure and sound
speed plummet; sound waves cease;
and atomic matter starts falling into
dark matter clumps. Photons move
freely, and we see them today as the
microwave background.

5–200 Myr...................................**The dark age**: By 5 Myr
($T \sim 600$ K), most photons are in the
infrared, and the Universe appears
dark. The atomic gas continues to
fall toward the dark matter clumps,
which grow more pronounced. Near
100 Myr ($T \sim 80$ K), the densest
clumps halt their expansion and
begin collapsing. By 200 Myr
($T \sim 50$ K), the first mini-halos of
dark matter form, and within these,
the atomic gas cools and collapses
to make the first stars whose light
brings the dark age to a close.

~200 Myr .... **First stars**: Because the first stars contain pristine gas, they are very massive, luminous, hot and short-lived (~1 Myr) compared to today's stars. They are quite strongly grouped because they form where peaks in density line up for all scales. They die in supernova explosions, liberating heavy elements that pollute the surrounding gas. Hopefully, the James Webb Space Telescope will soon see the first stars.

200–800 Myr............................ **Epoch of reionization**: The radiation from the first stars, and possibly the first quasars, ionizes much of the remaining neutral hydrogen and helium. A thin mist returns and partly obscures the microwave background. Hopefully, this epoch will soon be seen directly using low-frequency radio telescopes.

1–2 Byr.................................... **Infant galaxies**: Star groups merge to form infant galaxies. This time is characterized by frequent galaxy collisions, high star birth rates, and high supernova rates. Heavy elements production changes the pattern of star formation, making them of lower mass, less luminous, and longer-lived, like those of today.

2–3 Byr...........................	**Star birth and quasar peak**: In the dense environment of frequent galaxy collisions, the star birth rate reaches its maximum, as does the formation and feeding of supermassive black holes. Ever since, both these have been in decline.
~6 Byr	**First rich galaxy clusters**: Enough time has elapsed for the densest regions at the intersection of filaments to stop expanding and collapse. Preformed galaxies build a rich galaxy cluster, while gas falling in is heated to become the cluster atmosphere.
~7 Byr	**Deceleration/acceleration changeover**: Decelerating expansion ($S \propto t^{2/3}$) changes to accelerating expansion ($S \propto 2^t$), with a brief period of coasting ($S \propto t$) in between. This occurs before $\rho_m = \rho_{de}$ because for the same density, dark energy's tendency to fall outward is stronger than matter's tendency to fall inward.
~8 Byr	**First modern spiral galaxies**: Although some elliptical-like galaxies form in the first few Byr, classic spiral-like galaxies aren't seen until ~5 Byr ago. It takes time in an undisturbed environment for large spiral disks to develop.
~9 Byr	**Matter/dark energy equality**: At this time, the falling density of matter (dark and atomic) became equal to that of dark energy ($4.2\ m_H/m^3$).

9.1 Byr..**Sun and Earth form**: 4.6 Byr ago
the solar system forms in the disk of
the Milky Way. The Earth (and
other planets) form at almost the
same time as the Sun.

13.7 Byr ....**Today**: The cosmic age is currently
(as of 2008) known to 1% accuracy
(± 0.13 Byr).

+3–4 Byr ....**M31 and Milky Way begin
merging**: The M31 (Andromeda)
galaxy and the Milky Way galaxy
are falling together and will start to
merge in 3–4 Byr (the time is not
known accurately because the
sideways velocity isn't yet known).

+5 Byr ....**Sun dies**: The Sun goes through its
red giant phase (incinerating the
Earth's surface) and ultimately
ejects its atmosphere, becoming a
white dwarf.

~20 Byr ....**Growth of structure ceases**: By 20
Byr, $\Omega_{de} \sim 0.9$, so expansion is
almost pure exponential, shutting
down further development of large-
scale structure.

~100 Byr ....**Virgo cluster passes outside
event horizon**: Dark energy's
accelerating expansion creates an
event horizon at a distance of
~15 Bly (r_H for $t_H = 15$ Byr).
The Virgo Cluster crosses this in
about 100 Byr ($S \sim 270$) and
becomes invisible.

~1000 Byr ....**Longest-lived stars die**: The longest-lived stars use up their fuel and die in about 1 trillion years. How long stars continue to form is unclear and depends on whether fresh gas can cool and collapse. With such long timespans ahead, let's choose not to speculate further.

Glossary

absolute zero: The temperature at which all atomic and molecular motion ceases. This forms the zero for the Kelvin scale: $0 \text{ K} = -273.16°\text{C}$.

absorption line: A dark line appearing in a spectrum, caused by atoms or ions absorbing a particular color from a continuous (usually thermal) spectrum. The usual situation is a hot object (producing the thermal spectrum) with a cooler gas in front, as in a star's atmosphere.

accretion disk: A disk of gas rotating about a central object. As the gas slowly spirals inward, it releases gravitational energy. If the central object is compact (e.g., a black hole), the accretion disk gets very hot and luminous.

active galaxy: A galaxy whose central massive black hole is currently accreting gas and hence releasing energy, either as radiation or outflowing jets.

anisotropy: Not isotropic; has variations in different directions. In cosmology, the term is applied to the slight patchiness in the microwave background (hence "WMAP": Wilkinson Microwave *Anisotropy* Probe).

annihilation: Term used for when a particle and antiparticle collide and "annihilate." In fact, they are simply converted into energy, often photons.

anthropic reasoning (principle): A line of reasoning that uses the fact that we exist to explain something about the properties of the Universe. For example, the proton-to-neutron mass ratio must be close to its measured value; otherwise, atoms would not exist, and neither would we.

antimatter: An identical version of every subatomic matter particle, but having opposite quantum numbers (e.g., charge, parity, spin). When a particle meets its antiparticle, they convert to energy (usually photons).

apparent brightness: How bright an object appears to us in the sky. Its units are watts per square meter. It is related to the object's luminosity and distance by the famous inverse square law: $b = L/4\pi d^2$.

arc minute: Small unit of angle, 1/60 of a degree.

arc second: Very small unit of angle, 1/60 of an arc minute, 1/3600 of a degree.

astronomical unit (AU): Average distance from the Earth to the Sun: 149.6 million km, or 93 million miles.

atom: Basic unit of a chemical element; roughly 0.1 nm in size; composite structure with a small, heavy, positively charged nucleus surrounded by a cloud of orbiting, lightweight, negatively charged electrons.

atomic matter: The term we use in this course for 1 of the 5 basic cosmic constituents, comprising protons, neutrons, and electrons. The equivalent term used by astrophysicists is "baryonic matter."

atomic nucleus: Positively charged core of atom containing 99.98% of the mass; roughly 10^{-6} nm in size, or 100,000 times smaller than the atom.

baryon: A class of particles containing three quarks (e.g., protons, neutrons).

beryllium barrier: Refers to the fact that fusion reactions in the early Universe were prevented from making elements heavier than helium because a crucial stepping-stone was missing: Beryllium-8 is effectively unstable (lifetime 10^{-16} seconds). If it were stable, helium fusion could have passed on through beryllium to carbon, and so on, to heavier elements.

Big Bang: Term used to denote, generally, the "explosive" origin of the Universe. There is no clear definition: It can mean the hot early phase (minutes), or the preinflation hypothetical first moment, or the inflationary launching.

Big Crunch: Hypothetical collapse of the Universe, similar but opposite to the Big Bang. It might arise if the Universe's expansion stops and reverses.

billion: Nonscientific term for giga, that is, 10^9.

black hole: A region where matter is so compressed that it closes (warps) the local space-time, preventing even light from escaping. In this course, we are mainly interested in "supermassive" black holes (10^6–10^9 M_{Sun}) that are found at the centers of galaxies. The gravitational fields of black holes are so strong that a significant fraction of the matter falling into them can be converted into energy.

boson: A particle with integer spin. The fundamental bosons are carriers of the 4 forces of nature. This distinguishes them from the fundamental fermions with half-integer spin that are the matter particles.

causal contact: If 2 points are in causal contact, then forces moving at or below light speed can travel between the 2 points. In a Universe with a finite age, it is possible for regions sufficiently far apart never to have been causally connected. Inflation makes the size of these causally connected regions very much larger than in the standard radiation-dominated Hot Big Bang.

Cepheid variable star: A particular type of pulsating star (named after Delta-Cepheus) whose brightness varies in a periodic way. There is a tight correlation between the luminosity of the star and the period of its variation, which can be used to measure the distance to Cepheids in nearby galaxies.

classical physics: The branch of physics specifically dealing with things significantly larger than atoms. It is characterized by clean deterministic behavior, which differs from the more subtle probabilistic behavior of the subatomic realm that is treated using quantum physics.

closed Universe: A Universe with positive spatial curvature whose total volume is finite. In such a Universe, the average density is greater than the critical density.

CMB: *See* **cosmic microwave background**.

cold dark matter: Refers specifically to the kind of dark matter in which the particles are heavy (e.g., greater than a proton) and so move slowly. They can therefore clump up to provide the halos into which atomic matter can fall and make stars and galaxies. Observations suggest real dark matter is indeed "cold," and not "hot."

comoving viewpoint: A viewpoint that expands with the Universe. Movies of structure formation usually take this viewpoint, so you don't see the cosmic expansion.

compactification: Refers to the process by which higher dimensions curl up, possibly near the Planck time, leaving three large spatial dimensions with tiny Calabi-Yau spaces at each point. In String Theory, the shape of these spaces defines the laws of physics, because it is across these spaces that strings vibrate, making the suite of particles.

concordance model: The particular cosmological model that best matches all the observational data sets. It is a Freedman-Robertson-Walker model, with Euclidean geometry and component omegas of 4.4%, 23%, 73%, 5.0×10^{-3}%, 3.4×10^{-3}% for atomic and dark matter, dark energy, photons, and neutrinos. It has a primordial roughness spectrum close to $P(k) \propto k$, with amplitude matching the CMB patchiness, and a period of reionization near 400 million years. Its current Hubble constant is 22 km/s/Mly, and its current age is 13.7 billion years.

confinement: *See* **quark confinement**.

convective dredge-up: The process by which newly made atomic nuclei are brought up from a star's core to its surface in huge hot convection currents.

cosmic evolution: Systematic change over time (*not* a Darwinian process).

cosmic microwave background (CMB): Faint omnidirectional microwave radiation associated with the Big Bang. It comes from hot glowing gas near 380,000 years, which we see as a very distant wall of fog whose light has been redshifted into the microwave part of the spectrum. It is extremely uniform (showing that the Universe is homogeneous and isotropic on large scales), with slight patchiness (revealing roughness that will ultimately make stars and galaxies).

cosmic neutrino background (CNB): Analogous to the CMB, but for neutrinos. The time of last scattering is ~1 second. It won't be detected in the foreseeable future.

cosmological constant: The term introduced by Einstein in his field equations for gravity that provided a long range (cosmological) repulsive force to balance the attractive force from matter, allowing him to create a static Universe. He famously rejected this constant when Hubble later discovered cosmic expansion, but today it has been reintroduced in a slightly different form as a vacuum energy. Currently this is the favored explanation for dark energy.

cosmological principle: States that the Universe looks the same from all locations. In practice this means it is isotropic and homogeneous (smooth) on large scales (greater than 200 Mly).

cosmology: The scientific study of the origin and development of the Universe.

creation myth: An account of the origin of the Universe and our place in it, which has arisen primarily in a historical, cultural, and nonscientific manner.

critical density: Refers to that density that renders the cosmic space "flat" (Euclidean). Its value depends on the expansion rate: $\rho_c = 3H_0^2/8\pi G$, which gives 9.7×10^{-27} kg/m^3 = 5.8 m_H/m^3 if we use $H_0 = 22$ km/s/Mly.

current (comoving) distance: The distance to a galaxy today. In an expanding Universe, this is greater than either the distance when the light set out (the emission distance) or the distance the light itself traveled (the light-travel distance, also known as the look-back time). For example, the gas visible in the microwave background has a current distance of 46 Bly, an emission distance of 46 Mly, and a light-travel distance of 13.7 Bly.

curvature: In General Relativity, the 4-dimensional space-time is curved by mass and energy, and this curvature is manifest as the force of gravity. In a homogeneous and isotropic Universe, the time and space parts separate cleanly, and the space part alone exhibits simple uniform curvature, which gives rise to the departure from the normal Euclidean rules of geometry. *See also* **geometry**.

dark energy: The primary cosmic component that is thought to be a small energy associated with the vacuum of space and is therefore uniform throughout the Universe. The fact that it isn't diluted by cosmic expansion actually causes the expansion to accelerate. It plays the same role in cosmic expansion as Einstein's cosmological constant.

dark energy era: The time since 9 billion years, in which dark energy has dominated the cosmic density and caused the expansion to grow exponentially with time ($S \propto 2^t$).

dark matter: The dominant cosmic component of matter, roughly 6 times more abundant than atomic matter. It is thought to be a relic particle left over from the Big Bang that only responds to the weak force and gravity, and is hence "dark" (invisible).

decibel (dB): Measures the loudness of sound waves: 135, 75, and 15 dBs correspond to deafening, noisy, and quiet. It depends on the ratio of pressure variations to the average pressure: $\Delta P / P$.

decouple: When two previously interacting components no longer interact. For example, neutrinos decouple at 1 second, and photons decouple at 380,000 years.

density: Measured as (mass of object)/(volume of object), with units of kg/m^3. Thus, lead is denser than wood, which is denser than air.

deuterium: A version (isotope) of hydrogen with a neutron in addition to the single proton. Its creation in the minute-old Universe allows us to measure the total atomic matter content of the Universe. It is roughly 10^{-5} the abundance of normal hydrogen.

deuterium bottleneck: Refers to the fact that the fusion reactions in the early Universe couldn't proceed until the fragile deuterium nucleus could survive long enough to combine to make helium. The delay was about 100 seconds before the temperature was sufficiently low that photons no longer broke up the deuterium nuclei.

dipole: The yin-yang pattern seen in the CMB all-sky map caused by the Earth's motion relative to the CMB frame (~600 km/s toward Hydra).

dissipationless collapse: The gravitational collapse of a system in which *no* energy is lost. Such collapse usually undergoes a period of rather chaotic readjustment to make a rather low-density final system. Systems of stars, and dark matter, collapse in this way.

dissipative collapse: The gravitational collapse of a system in which energy is lost. Typically, energy is lost by radiation, and the collapse proceeds to high density.

distance ladder: Because no single method can measure distances to both near and far objects in the Universe, the set of overlapping methods, in which nearby ones calibrate more distant ones, is called the distance ladder. In these lectures we look at just 5 steps (rungs).

Doppler shift: The change in observed wavelength due to the relative motion of source and observer. In sound this corresponds to a change in pitch; in light it corresponds to a change in color. The fractional change in wavelength, $\Delta\lambda/\lambda$, is equal to the relative velocity to the velocity of the wave, V/c. Motion away gives a red shift while motion toward gives a blue shift.

electromagnetic force: One of the 4 fundamental forces. It is carried by the photon and acts between any charged particles: for example, the electrons and protons within atoms.

electromagnetic spectrum: The sequence of electromagnetic waves arranged from long to short wavelength: radio, infrared, light, ultraviolet, X-ray, gamma ray. These waves also correspond to photons of different energies and associated wavelengths.

electron: A subatomic particle with mass 9.1×10^{-31} kg (1/1839 proton mass), with negative charge. Electrons are fundamental particles with no interior particles.

electroweak force: The combined electromagnetic and weak fundamental forces. This occurs at energies above about 100 GeV, or before 0.1 nanoseconds.

element: A substance whose atoms have a specific number of protons. There are 92 natural elements, for example, hydrogen (1), helium (2), carbon (6), iron (26), uranium (92).

elliptical galaxy: One of the primary kinds of galaxies, with no disk or spiral arms, but rather a single smooth, spheroidal distribution of stars.

emission distance: The distance to an object when its light set out. In an expanding Universe, the distance to a remote galaxy can change while the light is traveling to us, so the distance to the galaxy when its light set out is less than its distance today (when the light arrives).

emission line: A bright line of single color appearing in a spectrum, made by atoms or ions in a hot gas. Such lines are extremely useful to astronomers because they contain a great deal of information about the object producing the light.

energy density: The amount of energy in a given volume. In cosmology, this usually includes all forms of energy, including mass, because they all contribute to gravity. In this course, we have usually expressed the energy density as an equivalent mass density (i.e., dividing by c^2), with units of kilograms per cubic meter, or hydrogen-atom masses per cubic meter, or for the early Universe, tons per cubic centimeter.

equality: The time when the relativistic (radiation plus neutrino) density equals the matter (dark plus atomic) density. It occurs near 57,000 years.

escape velocity: The critical velocity needed for any object (e.g., a rock) to be launched in order not to turn around and fall back down. For a primary object of mass M and radius R, the escape velocity is: $V_{esc} = \sqrt{(2GM/R)}$.

Euclidean geometry: The geometrical rules appropriate to flat surfaces or flat spaces, as specified by the Greek mathematician Euclid around 300 B.C.

event horizon: A boundary beyond which nothing can be seen either now or in the future. In cosmology, an event horizon forms if the expansion is accelerating. For the concordance model, its value today is 15 Bly. An event horizon also forms around a black hole, beyond which nothing can be seen (because space is flowing inward at light speed and accelerating).

expansion history: Refers to the relative size or scale of the Universe over cosmic history. In these lectures, we call it $S(t)$, the change of scale factor with time. It starts at 0 at the Big Bang and is 1 today. During the radiation, matter, and dark energy eras, it takes the form: $S \propto t^{1/2}$, $S \propto t^{2/3}$, and $S \propto 2^t$.

exponential axes: When the intervals along the axes of a graph correspond to jumps by a multiplicative factor, we say the axes are exponential (or logarithmic). For example: 10 to 100 to 1000, etc. These kinds of axes are extremely useful because they allow one to see very small and very large things together on the same graph.

family: The fundamental particles of the standard model seem to come in 3 repeating patterns (also called generations), of successively higher mass. The lightest include the electron, its neutrino, the up quark, and the down quark. Similar sets of 4 particles make up the other 2 families. Although Big Bang helium production shows no more families exist, it is still not known why there are just 3.

far-infrared: Electromagnetic radiation with wavelength 30–300 μm. Galaxies with strong far-infrared emission usually have a high rate of star birth.

fermion: A particle with half-integer spin. The fundamental fermions are the matter particles, such as quarks, electrons, and neutrinos. This distinguishes them from the fundamental bosons, which have integer spin and are the force-carrying particles.

flat (Euclidean) Universe: A Universe with zero spatial curvature, so the simple Euclidean rules of geometry apply. In such a Universe, the average density is equal to the critical density. (Notice, there is nothing "flat" or squashed about this Universe—the term refers to the nature of the 3-D spatial geometry).

flatness problem: Refers to the unreasonably fine tuning associated with producing an approximately flat Universe today. The young Universe would need to be launched exceedingly close to flat. Of course, inflation does just that.

Fourier transform: A mathematical technique for finding the amplitudes and phases of sine waves that make up any function. The technique is used in cosmology to recast the density variations $\Delta\rho/\langle\rho\rangle(r)$ as a power spectrum $P(k)$, where $k = 2\pi/\lambda$ is a spatial frequency that characterizes how "big" or "small" the component sine waves are. In these lectures we use the term "roughness spectrum" for $P(k)$.

frequency: The number of waves passing each second, or equivalently, the number of wave oscillations per second. The unit is Hertz, or "per second," or /sec.

Friedmann equation: An equation derived by Alexander Friedmann in 1922 from the Einstein General Relativity equations for an isotropic homogeneous Universe. In this course we only consider Friedmann's energy balance equation, though there is another (acceleration) equation. Solutions to these give the expansion history of the Universe. In this course, we write the equation as $(V/V_0)^2 - [\Omega_{m,0}/S + \Omega_{r,0}/S^2 + \Omega_{v,0}S^2] = 1 - \Omega_{tot,0}$, where the three omegas are for today's matter, radiation, and vacuum density; V/V_0 is the expansion speed relative to today's expansion speed; and S is the scale factor.

fundamental particle: One of the noncomposite subatomic particles. Examples are photon, electron, quarks, and gluons (*not* protons or neutrons, which contain 3 quarks).

fundamental tone: The lowest and strongest frequency produced by a musical instrument. The term can be used for the first peak in the CMB sound spectrum.

fusion: The combining of atomic nuclei to make heavier nuclei. If the collisions result from thermal motion, we call this thermonuclear fusion.

Gaia satellite: A European satellite due to be launched in 2011 that will measure parallaxes (and hence distances) to millions of stars throughout our galaxy.

galactic cannibalism: The process whereby a smaller galaxy falls into a larger galaxy, gets tidally shredded, and ultimately gets "consumed" by the larger galaxy.

galaxy: A basic cosmic unit containing between 10^7 and 10^{11} stars, with gas and dust, all bound by gravity. Components include nucleus, bulge, disk, and halo of dark matter. Three types are spiral, elliptical, and irregular. In total, about 10^{11} galaxies are visible.

galaxy cluster: The largest currently relaxed systems, containing up to several thousand galaxies with masses $\sim 10^{13}$–10^{15} M_{Sun}.

galaxy web: The network of filaments and sheets of galaxies that fill the Universe.

Gaussian random field: Refers to fluctuations whose phases (locations) are independent of each other. Inflation predicts fluctuations with this property.

General Relativity: Einstein's second (1915) theory of relativity, which concerns gravity, space, and time. It supersedes Newton's theory of gravity and has never failed any experimental test, even in contexts in which Newton's theory does.

geometry: In cosmology, this refers to the rules of geometry that apply to simple figures arranged in 3 dimensions. For example, if the circumference of a circle is more/equal/less than $2\pi r$, the geometry is said to be closed/flat/open. In general, the simple Euclidean (or flat) rules may not apply to spaces curved by mass or energy.

GeV: 10^9 eV (electron volts), a unit of energy. A gas in which particles have this energy has a temperature of about 10 trillion K (10^{13} K).

giga-: Prefix for 10^9, or a billion times; for example, giga-year (Gyr), giga-light-year (Gly).

gluon: A fundamental particle that carries the strong force, binding quarks together inside protons and neutrons.

Grand Unified Theory (GUT): Aims to unite the electromagnetic, weak, and strong forces into a single theory, with a single force. It is thought that this merging of the 3 forces occurs near 10^{-36} seconds and energy 10^{16} GeV or 10^{28} K. Most GUT theories predict that the proton should be very slightly unstable.

gravitational energy: The energy in the gravitational fields acting between masses. For 2 masses M and m a distance R apart, the gravitational energy is $-GMm/R$. The most important fact about gravitational energy is that it is negative: One must put energy into these objects to separate them to infinity where their energy is zero.

gravitational lens: The bending of light by a massive object tends to "image/distort" a background light source (e.g., clusters of galaxies imaging background galaxies). By measuring the amount of distortion (the strength of the lens), one can calculate the mass contained in the lens.

gravitational red shift: Photons lose energy when they move "uphill," away from a mass concentration. This causes a red shift.

gravitational waves: Fluctuations in the geometry of space that move at light speed and are made when masses accelerate. Merging black holes are strong sources of gravitational waves. It is hoped that gravity waves will be detected in the next decade or two.

graviton: A massless boson particle that is postulated to carry the force of gravity.

gravity: Fundamental attractive force between all forms of mass and energy. First described by Newton as an "inverse square law of attraction," later described by Einstein as the motion of objects across a space-time warped by matter. In a quantum description, gravitons (massless bosons) carry the force.

GUT: *See* **Grand Unified Theory**.

hadron: Particles that are made of quarks, such as protons, neutrons, and mesons, and their antiparticles.

hadron era: The brief period between about 10^{-5} and 10^{-4} seconds when hadrons dominated the cosmic density. The period began with quark confinement near 10 microseconds and ended with proton-antiproton annihilation near 100 microseconds.

halo: The large region surrounding a galaxy; contains mainly dark matter, a few stars, and a few dwarf companions. Halos are the primary structures formed when dark matter collapses and merges during cosmic expansion, and they provide the gravitational pockets within which galaxies form and reside.

Hammer projection: A way of showing the surface of a sphere as a flat oval. This is convenient when showing maps of the full 360° sky.

harmonic: An overtone with simple frequency ratio to the fundamental. Often caused by systems vibrating 2, 3, 4, or more times the mode of the fundamental. Harmonics are seen in the CMB sound spectrum, caused by multiple oscillations since the Big Bang.

helium: The second element in the period table, with atomic nuclei containing 2 protons and (usually) 2 neutrons. It is the second-most abundant element in the Universe (25% by mass) and was made in the minute-old Universe.

hertz: Unit of frequency (Hz); measured in number per second (e.g., 440 Hz is concert A).

Hertzsprung-Russell diagram: A graph with surface temperature (*x*-axis) and luminosity (*y*-axis), on which stars can be placed. Stars of different mass, age, and composition lie at different places. For a cluster, the pattern of stars can be used to measure the age of the cluster.

Higgs boson: A hypothetical particle with zero spin. It is the last particle in the standard model yet to be experimentally discovered. It plays an important role in the electroweak theory, and its field gives all the other particles their measured masses. The particle is named after Peter Higgs, who postulated its existence in 1964.

Hipparcos satellite: The European satellite that measured the parallax angles (and hence distances) for 120,000 stars with a precision of about 1/1000 arc second.

homogeneous: The same at all locations. Averaged over sufficiently large volumes (larger than about 200 Mly), the Universe is homogeneous.

horizon: General term used in cosmology to signify a boundary of vision or influence. Three important horizons are visibility horizons, event horizons, and Hubble distances. In general, horizons arise because of the finite speed of light, the finite time since the Big Bang and cosmic expansion.

horizon problem: The large-scale uniformity of the Universe cannot be explained by the Standard Hot Big Bang Theory. For example, 2 regions on the microwave background separated by more than 2° have never been in causal contact (they are outside each other's horizons). The problem arises because the Standard Big Bang is dominated by radiation, with an initial infinite expansion speed. Inflation can solve the problem because its expansion starts slowly and accelerates.

hot dark matter: A hypothetical kind of dark matter whose particles are so light that they move too fast to clump up to make galaxy halos (for example, neutrinos). Current observations suggest that real dark matter is primarily "cold," and little if any of it is hot.

HST Key Project: A major observational project using the Hubble Space Telescope to measure distances to 18 galaxies using Cepheid variable stars. The project was led by Wendy Freedman and completed in 1999, and it played a major role in measuring the Hubble constant.

Hubble constant: Measures the current rate of cosmic expansion. It gives the recession speed between 2 (idealized) galaxies 1 million light-years apart, and its value for today's Universe is $H_0 = 22 \pm 2$ km/s/Mly. At other times, its value is different, and it is written "H" (without the naught) and is called the Hubble parameter.

Hubble flow: The uniform cosmic expansion that carries galaxies away from each other at a speed proportional to their separation.

Hubble law: The mathematical law that describes the cosmic expansion (Hubble flow) at any given time. Separation speed is proportional to separation: $V = H \times d$, where V is the velocity of separation and d is the separation (distance) between two galaxies. H is the Hubble parameter, which today is H_0, the Hubble constant.

Hubble radius: $r_H = c/H$, the distance at which the expansion speed equals the speed of light, c. It is also the Hubble time, $t_H = 1/H_0$, multiplied by the speed of light, c, and is hence roughly the size of the visible Universe.

Hubble time: $t_H = 1/H$ is a simple estimate for the time since the expansion began, assuming a constant expansion speed. For $H_0 = 22$ km/s/Mly, $t_H = 13.5$ Byr, which is close to the true age (13.7 Byr) because the early deceleration and later acceleration stages roughly cancel out.

Hubble Ultra Deep Field: The deepest image yet taken (in 2004) by the Hubble Space Telescope, and successor to the Hubble Deep Field (1995). Exposures totaling over 400 orbits revealed 10,000 galaxies in a region smaller than 1/10 the diameter of the full Moon. Some galaxies are seen only 800 million years after the Big Bang.

hydrogen: The simplest element, whose atoms have a nucleus with a single proton orbited by a single electron. Hydrogen is the most common element in the Universe (about 75% by mass), though most of it is in ionized form (the protons don't have an orbiting electron). It is the primary fuel for stars and the difficulty of the proton-proton fusion reaction is the main reason stars have such long lifetimes. *See also* **deuterium**.

hydrogen 21-cm emission: Radio emission of 21-cm wavelength produced by atomic hydrogen. Emission occurs when the electron's spin direction reverses (flips).

inflation: A brief period in the early (pre-nanosecond) Universe in which the primary component is a dense vacuum, which "falls outward" extremely quickly and both creates and launches the contents of the current Universe.

Inflationary Hot Big Bang: Extension to the Standard Hot Big Bang Theory, including a brief period of "inflation" that launches the expansion and creates the contents.

inflaton field: The quantum field associated with inflation. The inflaton particle and its field are generic terms, because the specific particle or field that drives inflation are not yet known. *See also* **quantum field, scalar field**.

intergalactic medium (IGM): The exceedingly tenuous gas between galaxies. After the dark age, it was reionized by the first stars and quasars, and it has remained so ever since. It is thought to contain about 80% of the atomic matter components in the Universe.

interstellar medium (ISM): The tenuous gas and "dust" particles that exist in the space between stars within galaxies. Its *average* density is about 10^6 atom/m^3, with about 1 dust particle/km^3.

intracluster medium (ICM): The very tenuous hot gas that fills clusters of galaxies. Because of its high temperature, it emits X-rays and is seen by X-ray telescopes.

invariance principle: When interactions between particles are unaffected by a change in some aspect of the interaction, then the interaction is said to be "invariant" under the change, and this in turn leads to a conservation law. Invariance principles and their conservation laws lie at the heart of the laws of physics.

inverse square law: Anything that decreases in proportion to the square of its distance is said to obey an inverse square law. In this course, we meet brightness (objects get fainter when they are further away) and the gravitational force (gravity's pull is weaker on more distant objects). In equation form, these are: $b = L/4\pi d^2$ and $F = GMm/d^2$. The $1/d^2$ makes these inverse square laws (double the distance, decrease by 4; triple the distance, decrease by 9; etc.).

ionized: A condition in which one or more electrons are removed from atoms, usually by energetic photons or collisions.

irregular galaxy: One of the primary kinds of galaxies, with no clear disk or bulge, but rather an irregular distribution of stars and gas. It is thought that the first galaxies resembled present-day dwarf irregular galaxies.

isotope: Although the atoms of a given element all have the same number of protons, they can have different numbers of neutrons, thereby giving different isotopes of that element. For example, hydrogen has 2 stable isotopes: Normal hydrogen has no neutrons, and deuterium has 1 neutron.

isotropic: The same in all directions. Averaged over sufficiently large regions and distances, the Universe is isotropic.

Kelvin: A unit of temperature, equal in size to °C but with 0 K at −273.16°C. Also called "absolute zero," where all molecular motion ceases. Thus, water's freezing and boiling points are 273 K and 373 K, and humans are roughly 310 K.

kilo-: Prefix for 10^3, or a thousand times. For example, kilometer (km), kilogram (kg).

kilogram: A conventional unit of mass; equal to roughly 2.2 lb.

kilometer: A common unit of distance; equal to 10^3 m, or roughly 0.62 miles.

kinetic energy: The energy associated with an object's motion, equal to $\frac{1}{2}mV^2$, where m is the object's mass and V its speed (near the speed of light, the formula is different).

km/s: Kilometers per second, a common unit of speed used in astronomy; equal to roughly 2250 mph.

Large Hadron Collider (LHC): The new-generation proton-antiproton accelerator at CERN, near Geneva. It's designed to achieve a collision energy of 14 TeV. Such collisions reproduce conditions at $t \sim 10^{-14}$ s and $T \sim 10^{17}$ K (100 times hotter than the electroweak unification temperature).

large-scale structure: Is the term usually given to describe the 3-dimensional pattern of galaxies that includes voids, clusters, filaments, and sheets.

lepton: A lightweight fundamental particle that doesn't feel the strong force. The leptons of the Standard model are the electron, muon, and tau, and their associated neutrinos, and their antiparticles.

lepton era: The period between about 10^{-4} and 10 seconds, during which leptons dominated the cosmic density. The period ended when the electrons and positrons annihilated.

light cone: The path of light rays in a space-time diagram. These form a 45° cone in a static space and make a teardrop shape in a standard expanding cosmological space. The light cone includes all events (specific location and time) that we see when we look out into the Universe (loosely: recent events nearby, events in the past further away, and very ancient events in the very distant Universe).

light curve: The graph of the brightness of an object against time. For example, the drop of brightness with time for a supernova, or the up-and-down brightness of a Cepheid variable star.

light-travel distance: The distance light itself traveled from an object to us. In an expanding Universe, this distance is greater than the distance when the light set out (the emission distance) and is less than the current distance (when the light arrives). The light-travel distance (in light-years) is also equal to the look-back time (in years).

light-year: The distance light travels in one year (9.47×10^{12} km); our cosmic unit of distance. (Astronomers often use parsecs, where 1 pc = 3.26 ly).

LISA: Laser Interferometer Space Antenna. A proposed set of 3 satellites separated by 5 million km that are designed to detect gravitational waves.

Local Group: A relatively sparse collection of 3 primary galaxies (the Milky Way, M31, and M33) and about 30 smaller galaxies. The group will ultimately merge into a single galaxy, starting in 3 or 4 billion years with the collision of the Milky Way and M31.

look-back time: The length of time light was traveling from an object to us. This is also the period back in time that we witness the object. The look-back time is also equal to the object's distance in light-years: An object 100 light-years away we see as it was 100 years ago. More accurately called the "light-travel distance."

luminosity: The intrinsic power radiated by an object, usually measured in watts (or multiples of the Sun's luminosity: $L_{Sun} = 3.8 \times 10^{26}$ watts). As the light moves away from the object, it spreads out, and at a distance d it has a brightness $b = L/4\pi d^2$, measured in watts per square meter.

Madau plot: Any graph that shows the star birth history of the Universe has come to be called a Madau plot, after Piero Madau, who (with his coauthors) first attempted to construct such a plot in 1996 using data from the Hubble Deep Field.

Magellanic Clouds: Two dwarf galaxies (the small and large clouds) that are close to our own galaxy, roughly 150,000 light-years away. They are only easily visible from the Southern Hemisphere; named after explorer Ferdinand Magellan.

magic number: Refers to particular numbers of neutrons or protons within atomic nuclei that are unusually stable, analogous to the stable electron shells of the inert (noble) gases. These numbers are 2, 8, 20, 28, 50, 82, and 126. They play an important role in nucleosynthesis, where they act as points of buildup during the s and r process of neutron addition.

main sequence: The primary line of stars on an Hertzsprung-Russell diagram, from cool dim stars of low mass to luminous blue stars of high mass. Stars burning hydrogen in their cores lie on this line, with location depending on the star's mass. The length of the main sequence for a cluster of stars acts as a kind of clock that can measure the cluster's age.

mass-energy: The energy associated with the matter in an object. Its value is given by Einstein's famous formula $E = mc^2$ and is roughly 20 megatons per kilogram. This is sometimes called "rest energy," because it's the energy of an object when it is stationary (with no kinetic energy).

matter era: The period of time from 57,000 to 9 billion years, during which matter (atomic and dark) dominated the cosmic density, giving an expansion proportional to time to the 2/3 power ($S \propto t^{2/3}$).

mega-: Prefix for 10^6, or a million times; for example, mega-light-year (Mly), megawatt (MW).

megaton: A unit of energy usually used for nuclear weapons: 1 megaton is a large-city-destroying hydrogen bomb and is equivalent to 46 grams of matter converted to energy.

merging: Occurs when two galaxies collide and merge to become a single galaxy.

meson: A class of subatomic particles containing two quarks (a quark and an antiquark).

meter: Conventional unit of length; roughly 1 yard (more precisely, 39.37 inches).

MeV: 10^6 eV (electron Volts); a unit of energy. A gas in which particles have this energy has a temperature of about 10 billion K (10^{10} K).

micro-: Prefix for 10^{-6}, or one millionth; for example, micrometer (μm), microwatt (μW).

microwave background: *See* **cosmic microwave background**.

milli-: Prefix for 10^{-3}, or one thousandth; for example, millimeter (mm), milliwatt (mW).

Milky Way: The band of light crossing the sky that corresponds to the disk of our galaxy seen edge on. The term is also used to refer to our galaxy, the Milky Way galaxy.

monopole problem: The Standard Hot Big Bang Theory cannot explain the absence of magnetic monopole particles that were made at the end of the GUT era. The problem is solved by inflation that dilutes the monopoles enormously.

M-Theory: *See* **String Theory**.

multiverse: The hypothetical ensemble of Universes that may exist in addition to our own Universe, possibly generated by inflation.

nano-: Prefix for 10^{-9}, or billionth; for example, nanometer (nm), nanosecond (ns).

neutrino: One of the fundamental particles, with little or no mass; moves at or close to the speed of light. Neutrinos are 1 of the 5 cosmic components, and they make up about 0.0034%.

neutron: A subatomic particle similar to a proton but having no charge.

nuclear valley of stability: The line along the neutron-proton graph with the most tightly bound atomic nuclei. The valley starts along the line $N(p) = N(n)$ but curves away so that stable heavy nuclei have more neutrons than protons. The valley is a useful concept in nucleosynthesis, when one considers how reactions (particularly neutron addition) make heavier nuclei that ultimately settle to the bottom of the valley.

nucleosynthesis: The process by which new atomic nuclei are created, usually by the combination of smaller nuclei in "thermonuclear fusion."

nucleus: Of an atom: the tiny bundle of protons and neutrons around which electrons orbit, roughly 100,000 times further away. Atomic nuclei contain 99.98% of the mass of the atom. Of a galaxy: the central regions of a galaxy, often including a supermassive black hole.

open Universe: A Universe with negative spatial curvature whose total volume is infinite. In such a Universe, the average density is less than the critical density.

parallax: The apparent shift of a star's position when viewed from different parts of the Earth's orbit. The angle, together with the size of the Earth's orbit, allows one to calculate the distance to the star, using the method of triangulation.

parsec: A unit of distance equal to 3.26 light-years. A star whose parallax angle is 1 arc second is at a distance of 1 parsec. In this lecture course, we use light-years instead of parsecs as our unit of interstellar and intergalactic distances.

particle desert: The enormous interval of time (in an exponential sense) from the GUT transition to quark confinement, 10^{-36}–10^{-5} seconds, during which the standard model predicts there are no significant changes in the particle population.

perfect cosmological principle: Extends the simple cosmological principle (the Universe looks the same from all locations) to add "at all times." Hence, the Universe would be unchanging. This principle lies at the heart of the Steady-State Theory.

period-luminosity relation: The relation between the pulsation period of a Cepheid variable star and the star's intrinsic brightness (luminosity). This relation can be used to measure distances to Cepheids and is a crucial step in the distance ladder.

phase transition: A change in the orderliness of a system, such as melting or boiling. In cosmology, it refers to the change in the behavior of quantum fields as the Universe cools. It is possible that such a phase change was responsible for inflation.

photon: A quantum of electromagnetic energy, or colloquially, a particle of light. Photons of longer wavelength have less energy than photons of shorter wavelength. Photons of ultraviolet light have an energy roughly equal to the energy that binds the outer electrons in atoms.

photon-baryon gas: Name given to the pre-recombination gas, to emphasize that the light, electrons, and protons are all tied together by collisions, making a single gas.

photosphere: The visible boundary of a star, where most of its light originates. In cosmology, the cosmic photosphere is equivalent to the wall of fog that made the microwave background.

pico-: Prefix for 10^{-12}, or a trillionth; for example, picosecond (1 ps = 1/1000 nanosecond).

Planck satellite: The next mission to map the microwave background (launch date October 2008, with final data release in 2012). The map should have significantly higher resolution and sensitivity than the Wilkinson Microwave Anisotropy Probe satellite and also provide improved data on polarization.

Planck time: Roughly 10^{-43} seconds. Loosely speaking, one tick of Nature's clock; the smallest interval of smooth time. Extrapolating the Standard Big Bang backward, at this time, gravity and quantum effects become entwined, and all our known laws of physics don't work.

planet: Collection of matter sufficiently large to self-compress into roughly spherical form, but sufficiently small not to initiate any thermonuclear fusion in its center.

positron: Another name for an antielectron (the antimatter version of an electron).

post-Copernican: A perspective that assigns no special status to the Earth's location.

power spectrum: Written $P(k)$. In these lectures, $P(k)$ refers to the spectrum of pressure variations that make up the primordial sound (the sound spectrum) and the spectrum of density variations that make up the Universe's roughness (the roughness spectrum). In general, $P(k)$ gives the relative strength of sine waves of a given frequency ($k = 2\pi/\lambda$) that are present in a complex pattern; it is derived by Fourier analysis of the pattern.

proton: A subatomic particle of mass 1.67×10^{-27} kg, with positive charge, containing three quarks. The number in an atomic nucleus defines the chemical element.

quantum field: In Quantum Field Theory, every particle has an associated field—an infinitely extended aspect of the vacuum that embodies the properties of the particle. With sufficient energy, the field can create its particle and antiparticle. In some circumstances, even with no particles present, a quantum field can have finite energy. Such fields can act as a vacuum energy and drive accelerating expansion. *See also* **scalar field, inflaton field**.

quantum fluctuation: An inherent variation of a system's properties over short time intervals. During inflation, quantum fluctuations of the vacuum are thought to have generated the seed roughness from which all the stars and galaxies ultimately form.

quantum physics: The branch of physics that specifically treats the behavior of atomic or subatomic scale systems. It is an exceedingly successful theory.

quark: A fundamental particle that makes up protons and neutrons. There are 6 quarks and 6 antiquarks. They carry a "color charge" that binds them together. Protons and neutrons contain 3 quarks; mesons contain a quark-antiquark pair. ("Quark" rhymes with "Mark.")

quark confinement: The time near 10 microseconds (200 GeV, 3×10^{12} K) when quarks became bound within protons, neutrons, mesons, and their antiparticles.

quark era: The period before 10 microseconds when quarks dominated the cosmic density and the Universe was filled with a quark-gluon plasma.

quark-gluon plasma: A condition of matter above 3×10^{12} K (200 MeV) and near nuclear density in which protons and neutrons "dissolve" into a sea of free quarks and gluons. The Universe was a quark-gluon plasma before about 10 microseconds. Heavy ion accelerators have recently been able to create a quark-gluon plasma.

quasar: An extremely luminous active galaxy, that is, with an accreting supermassive black hole at its center.

quintessence: A hypothetical vacuum energy that changes with time. It is one of the possible candidates for dark energy.

radiation: The term is usually used to refer to any form of electromagnetic energy (photons) of whatever wavelength. In cosmology, it can also refer to the combination of photons and neutrinos, both of which dominate the expansion in the young Universe.

radiation era: The period between 10 seconds and 57,000 years, during which radiation (photons and neutrinos) dominated the cosmic density, leading to an expansion proportional to the square root of time ($S \propto t^{1/2}$).

radiation pressure: The force exerted by photons (light) when they strike a surface. In terrestrial contexts, it is negligible. In the acoustic era, it dominates the pressure in the dilute, foggy (but brilliantly glowing) gas.

radio galaxy: A rare kind of galaxy that emits significant amounts of radio radiation. They are a kind of "active galaxy" and were more common in the early Universe.

recombination: A relatively brief transition near 380,000 years in which ionized gas became atomic gas; in other words, the free electrons combined with the protons and helium nuclei to make atoms. As a result, the Universe turned transparent, and we see this transition as the microwave background.

red shift: The shift of a spectrum to the red (longer wavelength) direction. This can be due to the Doppler effect from simple motion; or due to gravity if the light is climbing "uphill" away from a mass concentration; or, in the cosmological context, due to the expansion of space that stretches the photon as it crosses the Universe.

redshift stretch factor (*rsf*): As light waves cross the Universe, they are stretched by cosmic expansion. The stretch in wavelength matches the cosmic stretch, so $rsf = \lambda_{obs}/\lambda_{emit} = S_{obs}/S_{emit} = 1/S_{emit}$, where S is the scale factor (which is 1 today). In this course, we use *rsf* rather than the redshift parameter $z = (\lambda_{obs} - \lambda_{emit})/\lambda_{obs}$, which is less directly related to the cosmic expansion ($z = rsf - 1$).

reheating: The transformation of the scalar field vacuum energy into radiation and particles. This occurs at the end of inflation, at the start of the Hot Big Bang.

relativistic: Refers to particles or objects moving close to the speed of light, so their total energy is much greater than their rest mass. In a relativistic gas, the particles have kinetic energy greater than their mass-energy.

Relativistic Heavy Ion Collider (RHIC): An accelerator at Brookhaven National Labs designed to collide heavy atomic nuclei, such as gold or lead. In this way, it is able to create a quark-gluon plasma similar to conditions near 10 microseconds after the Big Bang, when the temperature was about 10^{12} K.

relativity: Theories concerning the transformation of fundamental properties (space, time, mass, length) between different observers. Einstein published two such theories. Special Relativity (1905) concerns velocity. General Relativity (1915) concerns acceleration and gravity. The famous $E = mc^2$ belongs to Special Relativity.

rest mass: *See* **mass-energy**.

rotation curve: The graph of rotation speed (y-axis) against distance from the center of the system (x-axis). Rotation curves yield information on the mass distribution within the system, and their failure to drop at large distances in galaxies points to the presence of dark matter halos.

R-process: *Rapid* addition of neutrons in stellar nucleosynthesis, thought to occur during a core-collapse supernova explosion when there is a very high flux of neutrons near the collapsing core.

R-process elements: The particular set of elements made in the R-process. In particular, there are 3 peaks in the element abundance graph—near selenium-80, tellurium-130, and platinum-192—containing elements made in the R-process.

Sachs-Wolfe effect: The gravitational red shifts and blue shifts arising from light originating in a lumpy gravitational landscape.

scalar field: A quantum field associated with a spin-zero particle. In some circumstances, scalar fields can have nonzero vacuum energy, and so drive inflation. *See also* **quantum field**, **inflaton field**.

scale factor: Written "S" in this course, it is the relative size, or scale, of the Universe compared to today (when $S = 1$). One can think of S as the relative separation between any 2 regions at any time, compared to today. It changes over time (increasing as the Universe expands), so it is more generally written $S(t)$.

second: Conventional unit of time. There are roughly 31 million seconds in 1 year.

slow roll: A term used to describe the change of the scalar field during inflation. In order to have significant inflation, the field must "slowly roll" down the potential curve.

solar luminosity: Total power output of the Sun in light (photons): $L_{Sun} = 3.86 \times 10^{26}$ watts.

solar mass: Mass of the Sun: $M_{Sun} = 1.98 \times 10^{30}$ kg $= 300,000 \times M_{Earth}$.

space-time: The 4-dimensional coordinate grid that contains 3 space dimensions and 1 time dimension. Unlike Newtonian space and time, which are separate, the 4 coordinates are woven together in ways that depend on the distribution of mass and energy.

space-time diagram: Useful graph of space (x-axis) and time (y-axis) allowing one to trace the paths of objects as well as light rays in an expanding Universe.

spectrum: A graph of the intensity of light (or other electromagnetic radiation) against wavelength (or frequency).

spiral galaxy: One of the primary kinds of galaxy, with bulge and disk and spiral arms.

S-process: *Slow* addition of neutrons in stellar nucleosynthesis, thought to occur during the red giant stage of some stars when there is a weak flux of free neutrons in a shell of nuclear fusion.

S-process elements: The particular set of elements made in the S-process. In particular, there are 3 peaks in the element abundance graph—near zirconium-90, barium-138, and lead-208—containing elements made in the S-process.

standard model: Current most comprehensive theory of fundamental particles. It is exceedingly successful, but also known to be incomplete.

starburst galaxy: A galaxy that is undergoing a relatively brief period ($\sim 10^8$ years) of intense star birth (10–1000 stars born per year).

Steady-State Theory: The cosmological theory originally advocated by Hoyle, Bondi, and Gold that adopts the perfect cosmological principle, namely, that on large scales the Universe does not change with time—it is in a steady state. To offset the dilution caused by expansion, more matter is created to keep the density constant.

String Theory: A theory that views fundamental particles as tiny strings whose modes of vibration determine which particle they are. The strings vibrate within a compact higher-dimensional space that exists at each location in our normal 3-D space. The theory is viewed by many physicists as the next great step forward, in part because it includes a quantum picture of gravity. The most recent development of String Theory is called M-Theory.

strong force: One of the 4 fundamental forces. It is carried by gluons and binds quarks together inside protons and neutrons. Some of this force "leaks" outside the proton and neutron and allows these to be bound together inside atomic nuclei.

structure problem: The inability of the Standard Hot Big Bang Theory to generate the necessary seed roughness to later form stars and galaxies. A successful cosmological theory must explain the origin of cosmic roughness. This is achieved in the Inflationary Hot Big Bang Theory via the amplification of quantum fluctuations.

sub-horizon growth: The term refers to the way in which density variations that are smaller than the horizon size grow in amplitude because gravity pulls matter from low-density regions to high-density regions. This mode of growth takes the high-density regions up to turnaround and collapse to form the first stars and galaxies. During the radiation era, sub-horizon growth is significantly suppressed, and this leads to a final roughness spectrum with a peak near 200 Mly, which corresponds to the horizon size at the end of the radiation era.

supercluster: The largest structures in the Universe, comprising groups of galaxy groups and clusters. Sizes are up to 100 Mly. We inhabit the Virgo (Local) supercluster. Superclusters are still expanding but have been slowed down relative to the normal Hubble flow.

super-horizon growth: The term refers to the way in which density variations that are larger than the horizon size grow in amplitude because of differential expansion, namely that slightly denser regions expand more slowly than less-dense regions, thereby amplifying the density variations. This is the primary mode of growth that takes the miniscule fluctuations from inflation and renders them sufficiently large to bridge the gap to the formation of stars and galaxies.

supernova: A cataclysmic stellar explosion. The two most common types are the thermonuclear detonation of a white dwarf star (used to measure distances to remote galaxies) and the core collapse of a massive star.

supernova remnant: The expanding debris from a supernova explosion, which contains many of the elements made during the star's life and during the explosion.

Supersymmetry: The name of a theory beyond the standard model that postulates a deep symmetry between fermions and bosons. In this theory, all particles have "super-partner particles": for each fermion there is a new boson (e.g., electron/selectron, quark/squark, neutrino/sneutrino), and for every boson there is a new fermion (e.g., photon/photino, W/Wino, Z/Zino). The lightest supersymmetric particle is thought to be stable, with a mass in the range 10–100 TeV, and this may be the dark matter particle. String theories naturally include Supersymmetry.

symmetry principle: *See* **invariance principle**.

temperature: In all things, the microscopic particles (atoms, molecules, etc.) are in a constant blur of motion. Temperature (in degrees Kelvin) is proportional to the average kinetic energy of these particles. Using units suited to the early Universe, the average energy of particles in a gas with temperature 10 billion K is about 1 MeV.

tera-: Prefix for 10^{12}, or a trillion times; for example, TeV (1000 GeV).

TeV: 10^{12} eV (electron Volts); a unit of energy. A gas in which particles have this energy has a temperature of about 10^{16} K.

thermal spectrum: The spectrum of radiation produced by a hot object. It has a peak with a gentle drop-off to long wavelengths and a steeper drop-off to short wavelengths. The peak wavelength and intensity of a thermal spectrum both depend on the temperature. Physicists sometimes call a thermal spectrum a "black-body spectrum" or a "Planck spectrum."

thermonuclear fusion: The type of reaction in which atomic nuclei collide and combine (fuse), their collision motion arising from high temperature.

Thomson scattering: Process by which photons are scattered by free electrons. This is the primary source of fogginess in the ionized gas of the young Universe.

threshold temperature: The temperature above which the energy of particle collisions can create new particle-antiparticle pairs. For example, the threshold temperature for the electron is about 4 billion K, because at that temperature the collision energies are around ½ MeV, which is the mass-energy of an electron.

tidal interaction: When galaxies get close to each other, their near and far sides experience a different gravitational force. The difference is called a tidal force, and it leads to several classic patterns, such as tidal arms and tidal streams.

tilt: Refers to the *deviation* from a simple primordial roughness spectrum of $P(k) \propto k$. It is thought to arise because expansion slows down at the end of inflation.

triangulation: The method used to measure the distance to an object by observing its position relative to background objects from 2 (or more) vantage points. The angular shift in the object position, together with the separation of the viewpoints, allows the distance to the object to be calculated.

Tully-Fisher relation: A tight correlation between galaxy rotation speed and galaxy luminosity; used to measure the distance to relatively nearby galaxies.

Universe: Everything there is, including space and time. Recently, it is restricted to our particular inflated space-time, with others belonging to the "multiverse."

vacuum: Refers to a region where no particles are present (including light): loosely speaking, pure emptiness. In modern physics, however, all of space is filled with quantum fields that continuously spawn virtual particle-antiparticle pairs. Thus, the modern concept of a vacuum is more than just "nothing."

vacuum energy: A finite energy associated with space itself, even when no particles are present. This situation can arise for certain types of quantum (scalar) fields. When this occurs, the vacuum generates its own gravity, and this in turn affects cosmic expansion, causing the Universe to fall outward at an accelerating pace. A small vacuum energy is the current favored explanation for dark energy.

valley of stability: *See* **nuclear valley of stability**.

virtual particle: Particles (and their antiparticles) constantly form and dissolve in and out of the vacuum. These are called virtual particles because they must return to the vacuum. They can be made real by injecting sufficient energy to provide their rest mass.

visibility horizon: Current limiting visible distance, where the look-back time is equal to the cosmic age. For the concordance model it is 46 Gly.

visible Universe: That part of the Universe currently visible to us. Loosely: out to 13.7 Bly distance. More precisely: our past light cone, terminating in the Big Bang.

void: A large (20–70 Mly) region with relatively few galaxies present. These have grown from slightly lower-density regions in the early Universe.

wavelength: The distance between successive crests or troughs of a wave.

weak force: One of the 4 fundamental forces. It is carried by the W and Z bosons and has such short range that it only operates inside particles. It is responsible for certain kinds of particle reactions, including some kinds of radioactivity.

weak lensing: Refers to a low-level gravitational distortion affecting the appearance of distant galaxies by foreground matter. Usually, the distortion amounts to a slight elongation and/or rotation of the galaxy image and is not noticeable for a single galaxy, but only when averaged over many galaxies. The phenomenon can be used to map out the distribution of dark matter, to show its evolving weblike pattern.

white dwarf: The burned-out core of a star, about Earth-sized but with the mass of the Sun, made from extremely compressed carbon and oxygen. If a white dwarf is in a binary system and the other star dumps matter onto it, the white dwarf can undergo thermonuclear detonation as a supernova explosion (type 1a). Such supernovae are used to measure distances to extremely remote galaxies.

WIMPS: Weakly interacting massive particles; a generic term for the class of particles that are thought to comprise (cold) dark matter.

worldview: The mental picture of the overall framework of reality, including physical, human, and possibly spiritual components.

w-parameter: This describes how the density of dark energy changes with expansion. For $w = -1$ there is no change, but for w slightly more (less) than -1, dark energy decreases (increases) with expansion. The technically correct name for w is the "equation-of-state" parameter, and the density obeys the relation $\rho_{de} \propto 1/S^{3(1+w)}$.

Biographical Notes

These biographical notes are only intended to highlight the achievements mentioned in the lectures (referenced as, e.g., L 6, 9) and do not include a broader account of the person's life and work.

Alpher, Ralph (1921–2007): American physicist who worked with George Gamow and Robert Herman on the physics of the Big Bang during 1948–1953, including the synthesis of chemical elements and the residual radiation from the initial fireball. In retrospect this work wasn't given the attention, or the credit, it deserved. (L 13, 25)

Aristotle (384–322 B.C.): Greek philosopher who advocated the geocentric paradigm, in part because if the Earth orbited the Sun, then the lack of observed stellar parallax would imply an unreasonably (to him) large distance to the stars. (L 1, 6)

Bessell, Friedrich (1784–1846): German mathematician and astronomer who was the first to measure (in 1838) the parallax shift of a star (61 Cygni) due to the Earth's orbital motion, giving a distance of about 3 parsecs, or 10 light-years. (L 6)

Birkhoff, George (1884–1944): American mathematician who showed that in General Relativity if the region surrounding a sphere was isotropic, one could ignore it and treat the sphere as isolated. Because the Universe is isotropic on large scales, Birkhoff's Theorem justifies using a Newtonian treatment for many aspects of cosmology (for example, deriving a simplified version of the Friedmann equation). (L 11)

Bondi, Hermann (1919–2005): Austrian-British physicist and mathematician who, with Thomas Gold and Fred Hoyle, developed the Steady-State Theory of cosmology in 1948. (L 3)

Burbidge, Geoffrey (b. 1925): British astronomer who helped map out, with Margaret Burbidge, William Fowler, and Fred Hoyle, many of the ways in which the chemical elements are made inside stars. Their work was published in a famous 1957 paper now referred to as B^2FH for its authors' initials. (L 24)

Burbidge, Margaret (b. 1919): British astronomer. *See also* **Burbidge, Geoffrey**. (L 24)

Cassini, Giovanni (1625–1712): Italian-French astronomer and mathematician who first measured with Jean Richer the geocentric parallax of Mars in 1672, which then allowed the size of the Earth's orbit to be determined. (L 6)

Copernicus, Nicolaus (1473–1543): Polish astronomer who famously published *De Revolutionibus Orbium Coelestium* (*On the Revolutions of the Celestial Spheres*) in the year of his death—the first clearly argued heliocentric cosmology. Because this removed the Earth from its central locality, we use the term "post-Copernican" to refer to any perspective in which we occupy an unremarkable position. The cosmological principle is an example of a post-Copernican idea. (L 3)

Curtis, Heber (1872–1942): American astronomer who publicly debated in 1920 against Harlow Shapley in Washington, DC, on the nature of the spiral nebulae. Although Curtis lost the debate, he was correct: The spiral nebulae are external galaxies. (L 6)

De Sitter, Willem (1872–1934): Dutch astronomer who derived a solution to Einstein's field equations in 1917 for a Universe containing only cosmological constant (vacuum), leading to exponential acceleration. On the basis of this, he predicted the phenomenon of red shift. In 1932 he published a flat pure matter cosmological model with Einstein: the Einstein–de Sitter Universe with $\Omega_m = \Omega_{tot} = 1$. (L 12)

Dicke, Robert (1916–1997): American astronomer and physicist who worked with Jim Peebles in the early 1960s to predict the microwave background. He also attempted to detect it but was famously scooped by Penzias and Wilson in 1964. (L 13)

Doppler, Christian (1803–1853): Austrian physicist and mathematician who, in 1842, studied the apparent change of pitch (frequency, wavelength) when a source of sound moved relative to an observer. The effect also occurs with light, giving rise to red and blue shifts for (very) fast-moving objects. (L 7)

Einstein, Albert (1879–1955): German American physicist whose work helped transform 19th-century classical physics into 20th-century modern physics. From 1905 to 1917, he created the Special and General Theories of Relativity and introduced the photon as a quantum particle of light. His most famous relation, $E = mc^2$, says that matter and energy are deep down the same thing. (L 4, 9, 10, 11, 12, 26, 27, 32, 35, 36)

Eratosthenes (c. 276–194 B.C.): Greek scholar who first measured the circumference of the Earth by comparing the elevation of the Sun on the same (midsummer) day in two cities. (L 10)

Euclid of Alexandria (~300 B.C.): Greek mathematician who wrote a famous text called *Elements*, deriving the rules of (flat) geometry from a small number of axioms. In the 19th century, Euclid's 5th axiom (parallel lines never meet) was dropped, giving "non-Euclidean geometries." In the early 20th century, General Relativity made use of these curved geometries to describe spaces curved by mass and energy. (L 10)

Fourier, Joseph (1768–1830): French mathematician who developed in 1822 the mathematical idea that any graph can be made from the sum of many sine waves of different wavelengths. His techniques are used in many areas of physics and engineering, and they provide a framework for describing roughness in the Universe. (L 17)

Fowler, William (1911–1995): American astrophysicist. *See also* **Burbidge, Geoffrey**. (L 24)

Freedman, Wendy (b. 1957): American astronomer who led the Hubble Space Telescope Key Project to measure distances to nearby galaxies using Cepheid variable stars and hence derived an accurate value for the Hubble constant. The project was completed in 1999, giving a value of 22 km/s/Mly with about 10% uncertainty. (L 6)

Friedmann, Alexander (1888–1925): Russian mathematician and cosmologist who used General Relativity in 1922 to derive 2 equations (the Friedmann equations) that describe how a homogeneous isotropic Universe expands or contracts over time. (L 11, 12)

Gold, Thomas (1920–2004): Austrian American physicist. *See also* **Bondi, Hermann**. (L 3)

Guth, Alan (b. 1947): American physicist who first proposed in 1980 the mechanism of inflation, which solved a number of problems with the Standard Hot Big Bang Theory. Guth's original theory was soon found to have serious problems of its own, but it initiated an entire field of research. (L 31)

Hertzsprung, Ejnar (1873–1967): Danish astronomer who published the first color-magnitude diagram for star clusters in 1913. Hertzsprung-Russell diagrams give great insight into the processes of stellar evolution and allow the determination of the ages of star clusters, including the oldest star clusters. (L 7)

Higgs, Peter (b. 1929): British theoretical physicist who proposed a mechanism in 1964 by which the electroweak symmetry was broken and the elementary particles acquired mass by interacting with a field (the Higgs field). The particle associated with this field is called the Higgs boson and is the only particle of the standard model not yet detected. (L 27, 29)

Hoyle, Fred (1915–2001): British cosmologist famous for his advocacy of the Steady-State Theory in the 1950s and 1960s. In 1950 he coined the graceless term "Big Bang" during some public lectures on the BBC. He was also responsible for much of the theory of the origin of the chemical elements, including the famous 1957 B^2FH paper with the Burbidges and Fowler. (L 3, 24)

Hubble, Edwin (1889–1953): American astronomer famous for at least 3 major contributions: his 1923 discovery, using Cepheids, that spiral nebulae were separate galaxies external to the Milky Way; his 1926 system of classifying galaxies based on their appearance; and his 1929 discovery of the increase of red shift with distance—the Hubble law, whose gradient is the famous Hubble constant. (L 6, 7)

Humason, Milton (1891–1972): American astronomer who worked with Hubble at the Mt. Wilson observatory. He began his unusual career in astronomy as a mule driver and janitor for the observatory, then rose via telescope assistant to staff member, ultimately making many of the observations credited to other astronomers, including Hubble. (L 7)

Leavitt, Henrietta (1868–1921): American astronomer working at the Harvard College Observatory who discovered in 1922 the period-luminosity relation for Cepheid stars in the Magellanic Clouds. Once calibrated, this relation provides a way to measure distances to nearby galaxies. (L 6)

Lemaître, Abbé Georges (1894–1966): Belgian cosmologist and Jesuit priest who first suggested in 1931 that cosmic expansion might have started in a singular moment, a "day without yesterday," to use his term. His theory suggested (incorrectly) that the beginning resembled a giant atomic nucleus that fragmented to make all the chemical elements. (L 25)

Linde, Andrei (b. 1948): Russian cosmologist, now at Stanford, who has played an important role in developing theories of inflation—in particular theories that address the possibility of a multiverse that arises from quantum effects either before or during inflation. (L 33)

Mather, John (b. 1946): American physicist responsible for the Cosmic Background Explorer satellite's 1992 accurate measurement of the thermal shape of the microwave background spectrum, for which he was awarded the 2006 Nobel Prize (with George Smoot). (L 13)

Maxwell, James Clerk (1831–1879): Scottish physicist most famous for unifying theories of electricity, magnetism, and light into a single electromagnetic theory, which views light as a small part of the electromagnetic spectrum. Maxwell also studied the thermal motion of atoms and molecules, and calculated their distribution of speeds. (L 4)

McCrea, William (1904–1999): Irish mathematician and cosmologist who showed, in 1934 with Arthur Milne, that because of isotropy and homogeneity, a Newtonian approach could yield considerable insight into many cosmological questions. (L 11)

Milne, Arthur (1896–1950): British astrophysicist. *See also* **McCrea, William**. (L 11)

Newton, Isaac (1642–1727): English scientist and mathematician. He formulated 3 laws of motion and the inverse square law of gravity, laying the foundations for classical mechanics. He recognized that an infinite Universe would clump up under gravity and suggested this might have been how stars formed. His concept of separate time and (Euclidean) space are almost always adequate, though distinct from Einstein's relativistic description of curved space-time. His 1687 treatise, *Principia*, is arguably the most influential scientific publication ever. (L 9, 11, 12, 17, 20, 26, 30, 33, 36)

Noether, Emmy (1882–1935): German mathematician whose principal area of study was abstract algebra. Her 1918 Noether's Theorem showed that a symmetry in the behavior of Nature leads to an associated conservation law. This is now a fundamental part of modern theoretical physics. (L 29)

Payne-Gaposchkin, Cecilia (1900–1979): English American astronomer who studied what influences the relative strength of absorption lines in stellar spectra, allowing chemical abundances in stars to be measured. This work constituted, in 1925, the first Ph.D. awarded at the Harvard College Observatory. (L 24)

Peebles, James (b. 1935): Canadian American cosmologist who investigated the properties of the hot early Universe with Robert Dicke in the early 1960s. He also helped develop the theory of how cosmic structures grow. (L 13, 16)

Penzias, Arno (b. 1933): German American physicist who worked at Bell Labs with Robert Wilson in the early 1960s. When developing a sensitive receiver for the horn antenna, they accidentally discovered the microwave background, for which they were awarded the Nobel Prize in 1978. (L 13)

Perlmutter, Saul (b. 1959): American astronomer who led the Supernova Cosmology Project that codiscovered in 1998 (with the High-z Supernova Search Team) the acceleration of the Universe's expansion. (L 9, 35)

Planck, Max (1858–1947): German physicist who played a major role in the foundation of quantum mechanics. His name is associated with the earliest cosmic epoch, the Planck time, when gravity takes on quantum form; his name has also been given to the upcoming European satellite to study the microwave background. (L 16, 29, 32, 33, 35)

Robertson, Howard (1903–1961): American mathematician and physicist who used the cosmological principle to elucidate the form of the cosmic space-time in 1935: a uniformly expanding or contracting curved space, and a separate cosmic time. This was independently derived by Arthur Walker in 1936. (L 10, 12)

Rubin, Vera (b. 1928): American astronomer most famous for measuring, in the early 1970s, the rotation curves of galaxies, finding them to be flat out to the visible limit of the galaxy, suggesting the presence of dark matter. (L 9)

Russell, Henry Norris (1877–1957): American astronomer who combined parallax measurements, spectra, and photometry of field stars to generate in 1914 a spectral type–luminosity diagram. Such diagrams are now called Hertzsprung-Russell diagrams and give valuable insights into stellar evolution, and they provide a way to measure the ages of star clusters. *See also* **Hertzsprung, Ejnar**. (L 7)

Ryle, Sir Martin (1918–1984): English radio astronomer who used the number of faint radio sources to test whether their population changes over time. After initial confusion, by the mid-1960s the results confirmed an evolving population, counter to the predictions of the Steady-State Theory. (L 3)

Sakharov, Andrei (1921–1989): Soviet physicist, cosmologist, and dissident who first defined in 1967 the conditions needed to generate a matter-antimatter asymmetry in the very early Universe. He also studied the growth of sound waves in the pre-recombination Universe. (L 16, 29)

Schmidt, Brian (b. 1967): American astronomer who led the High-z Supernova Search Team that codiscovered in 1998 (with the Supernova Cosmology Project) the acceleration of the Universe's expansion. (L 9, 35)

Shapley, Harlow (1885–1972): American astronomer who used Cepheid distance estimates to globular clusters to measure the size of the Milky Way galaxy, placing the solar system away from the galaxy center. He famously debated the external nature of the spiral nebulae with Heber Curtis in 1920 in Washington, DC, winning the debate but supporting the incorrect position. (L 6)

Slipher, Vesto (1875–1969): American astronomer working at the Lowell Observatory, who by 1917 had measured Doppler shifts of 25 spiral nebulae, finding large velocities, most of which were positive (moving away). Many of the red shifts used by Hubble in his 1929 discovery of cosmic expansion were measured by Slipher. (L 7)

Smoot, George (b. 1945): American physicist responsible for the COBE (Cosmic Background Explorer) satellite's 1992 measurement of the anisotropy of the microwave background, for which he was awarded the 2006 Nobel Prize (with John Mather). (L 13)

Starobinsky, Alexei (b. 1950): Russian theoretical physicist who coproposed inflation as a possible launch for cosmic expansion in 1979, independently (and somewhat differently) from Alan Guth. (L 31)

Toomre, Alar (b. 1937): American astrophysicist who first performed computer simulations in the early 1970s of interacting galaxies, discovering the origin of the classic tidal arms and bridges often seen in real colliding galaxies. (L 20)

Vilenkin, Alexander (b. 1949): Russian theoretical physicist, now at Tufts University, who first proposed eternal inflation and the possibility of a quantum origin of inflation from nothing. (L 33)

Walker, Arthur (1909–2001): British mathematician who derived in 1936 the form of the cosmic space-time for an isotropic, homogeneous Universe. *See also* **Robertson, Howard**. (L 10, 12)

Webb, James (1906–1992): American administrator of NASA during the Apollo period (1961–1968). He fought hard to maintain funding for NASA's manned program to land a man on the Moon, as well as for science programs in general. Hubble's successor telescope is named after him: the James Webb Space Telescope. (L 8, 19, 35)

Wilkinson, David (1935–2002): American physicist working at Princeton on studies of the microwave background, starting in 1965 with Robert Dicke, but later on the COBE (Cosmic Background Explorer) and Wilkinson Microwave Anisotropy Probe teams. (L 13)

Wilson, Robert (b. 1933): American physicist who won the Nobel Prize with Arno Penzias for discovering the microwave background in 1964 using the horn antenna at Bell Labs in New Jersey. (L 13)

Zeldovich, Yakov (1914–1987): Soviet physicist who worked in many areas, including the theory of the hot early Universe, the growth of structure, and galaxy formation. (L 13, 16)

Zwicky, Fritz. (1898–1974): Swiss American astronomer based at Caltech who proposed a number of important astrophysical ideas in the 1930s and 1940s, many of which turned out to be true, though not all. In our subject, he was first to infer the presence of dark matter in 1933 by measuring galaxy velocities in the Coma Cluster. (L 9)

Bibliography

There are many books on cosmology, at all levels. Here is a selection that overlaps with the material and level of this course. I have chosen 4 as primary texts: 2 are undergraduate textbooks written for nonscience majors, and 2 are written for more general readership. If you were to buy just 1 book, then I'd suggest Silk's if you want a lighter read, and Hawley and Holcomb's if you want a more systematic resource.

The other books emphasize a wide range of themes. Although some give a broad overview of cosmology (Barrow, 1994; Hogan; Rees, 2001; Silk, 2002), others place more emphasis on particular themes: historical (Alpher and Herman; Bartusiak; Fergusson; Kragh; Singh); CMB results (Lemonick; Mather and Boslough; Smoot and Davidson); dark energy (Goldsmith; Kirshner; Livio; Nicolson); dark matter (Rubin; Nicolson; Freeman and McNamara); inflation and the early Universe (Barrow, 2002; Guth; Weinberg, 1993; Vilenkin); fundamental physics (Greene; Susskind; Weinberg, 1992); black holes (Begelman and Rees); anthropic reasoning (Gribben and Rees; Rees, 1998, 2000; Susskind); geometry and curvature (Osserman); and the cosmologists themselves (Lightman and Brawer; Overbye; Rubin).

I have also included a few texts at a more advanced level for those of you who already have some background in astronomy and/or physics and want a deeper account of the field.

Finally, I've added some websites of interest, though as always, their continued existence isn't guaranteed.

Primary texts

Ferris, Timothy. *The Whole Shebang: A State-of-the-Universe(s) Report*. New York: Simon & Schuster, 1997. A very readable account of modern cosmology, though it predates some of the most recent results.

Hawley, John, and Katherine Holcomb. *Foundations of Modern Cosmology*. 2nd ed. New York: Oxford University Press, 2005. A comprehensive undergraduate textbook (for nonscience majors).

Jones, Mark, and Robert Lambourne. *An Introduction to Galaxies and Cosmology*. Cambridge: Open University Press, 2003. A textbook designed for the UK's Open (adult learning) University.

Silk, Joseph. *The Infinite Cosmos: Questions from the Frontiers of Cosmology*. Oxford: Oxford University Press, 2006. A relatively brief tour of almost all aspects of modern cosmology.

Recommended reading at a general level

Alpher, Ralph, and Robert Herman. *Genesis of the Big Bang*. Oxford: Oxford University Press, 2001.

Barrow, John. *The Book of Nothing: Vacuums, Voids, and the Latest Ideas about the Origins of the Universe*. New York: Vintage Books, 2002.

———. *The Origin of the Universe*. New York: Basic Books, 1994.

Bartusiak, Marcia. *Archives of the Universe: A Treasury of Astronomy's Historic Works of Discovery*. New York: Pantheon Books, 2004.

Begelman, Mitchell, and Martin Rees. *Gravity's Fatal Attraction: Black Holes in the Universe*. New York: W. H. Freeman, 1998.

Dressler, Alan. *Voyage to the Great Attractor: Exploring Intergalactic Space*. New York: Knopf, 1994.

Eliade, Mircea. *Myth and Reality*. Long Grove, IL: Waveland Press, 1998.

Fergusson, Kitty. *Measuring the Universe: Our Historic Quest to Chart the Horizons of Space and Time*. New York: Walker & Co., 2000.

Freeman, Ken, and Geoff McNamara. *In Search of Dark Matter*. New York: Springer Praxis Books, 2006.

Goldsmith, Donald. *The Runaway Universe: The Race to Find the Future of the Cosmos*. Cambridge, MA: Perseus, 2000.

Greene, Brian. *The Fabric of the Cosmos: Space, Time, and the Texture of Reality*. New York: Knopf, 2004.

Gribbin, John, and Martin Rees. *Cosmic Coincidences: Dark Matter, Mankind and Anthropic Cosmology*. New York: Bantam, 1989.

Guth, Alan. *The Inflationary Universe*. Cambridge, MA: Perseus, 1998.

Hogan, Craig. *The Little Book of the Big Bang: A Cosmic Primer*. New York: Copernicus, 1998.

Hu, Wayne, and Martin White. "The Cosmic Symphony." *Scientific American*, February 2004.

Kirshner, Robert. *The Extravagant Universe: Exploding Stars, Dark Energy, and the Accelerating Cosmos*. Princeton, NJ: Princeton University Press, 2002.

Kragh, Helge. *Conceptions of Cosmos: From Myths to the Accelerating Universe: A History of Cosmology*. Oxford: Oxford University Press, 2007.

———. *Cosmology and Controversy: The Historical Development of Two Theories of the Universe*. Princeton, NJ: Princeton University Press, 1996.

Lemonick, Michael. *Echo of the Big Bang*. Princeton, NJ: Princeton University Press, 2003. An account of the Wilkinson Microwave Anisotropy Probe mission.

Lightman, Alan, and Roberta Brawer. *Origins: The Lives and Worlds of Modern Cosmologists*. Cambridge, MA: Harvard University Press, 1990.

Livio, Mario. *The Accelerating Universe: Infinite Expansion, the Cosmological Constant, and the Beauty of the Cosmos*. New York: Wiley, 2000.

Mather, John, and John Boslough. *The Very First Light: A Scientific Journey Back to the Dawn of the Universe*. New York: Basic Books, 1998.

Nicolson, Ian. *The Dark Side of the Universe: Dark Matter, Dark Energy, and the Fate of the Cosmos*. Baltimore, MD: Johns Hopkins University Press, 2007.

Osserman, Robert. *Poetry of the Universe*. Garden City, NY: Anchor, 1996. Explores the nature of curved space.

Overbye, Dennis. *Lonely Hearts of the Cosmos: The Story of the Scientific Quest for the Secret of the Universe*. Boston: Back Bay Books, 1999.

Rees, Martin. *Before the Beginning: Our Universe and Others*. Cambridge MA: Perseus Books Group, 1998.

———. *Just Six Numbers: The Deep Forces That Shape the Universe*. New York: Basic Books, 2000.

———. *Our Cosmic Habitat*. Princeton, NJ: Princeton University Press, 2001.

Rubin, Vera. *Bright Galaxies, Dark Matters*. New York: AIP Press, 1996.

Singh, Simon. *Big Bang: The Origin of the Universe*. New York: HarperCollins, 2004.

Smoot, George, and Keay Davidson. *Wrinkles in Time: Witness to the Birth of the Universe*. New York: HarperCollins, 1994.

Susskind, Leonard. *The Cosmic Landscape: String Theory and the Illusion of Intelligent Design*. New York: Little, Brown and Company, 2006.

Vilenkin, Alex. *Many Worlds in One: The Search for Other Universes*. New York: Hill and Wang, 2006.

Weinberg, Steven. *Dreams of a Final Theory: The Search for the Fundamental Laws of Nature*. New York: Pantheon, 1992.

————. *The First Three Minutes: A Modern View of the Origin of the Universe*. 2nd ed. New York: Basic Books, 1993.

Recommended reading at a more advanced level

Keel, William. *The Road to Galaxy Formation*. Chichester, UK: Springer Praxis, 2002. A graduate text that emphasizes observational work on galaxies and galaxy evolution.

Linder, Eric. *First Principles of Cosmology*. Harlow, UK: Addison-Wesley-Longman, 1997. A concise text at the advanced-undergraduate/graduate level.

Longair, Malcolm. *The Cosmic Century: A History of Astrophysics and Cosmology*. Cambridge: Cambridge University Press, 2006. A masterly review of 20th-century research in astrophysics and cosmology.

————. *Galaxy Formation*. 2nd ed. Heidelberg, Germany: Springer, 2008. A classic graduate text that emphasizes theory over observations.

Pagel, Bernard. *Nucleosynthesis and Chemical Evolution of Galaxies*. Cambridge: Cambridge University Press, 1997. A graduate text on the origin of the elements and their distribution within galaxies.

Rees, Martin. *New Perspectives in Astrophysical Cosmology*. 2nd ed. Cambridge: Cambridge University Press, 2000. A slim volume that discusses ongoing problems and questions in cosmology (originally written in 1995).

Ryden, Barbara. *Introduction to Cosmology*. San Francisco: Pearson Education, Inc., 2003. An undergraduate (for science majors) text.

Schneider, Peter. *Extragalactic Astronomy and Cosmology: An Introduction*. Heidelberg, Germany: Springer, 2006. A comprehensive advanced-undergraduate/graduate text with both theoretical and observational material.

Silk, Joseph. *The Big Bang*. 3rd ed. New York: Henry Holt and Company, 2002. A book for general readership, but pitched at a somewhat higher level than most. Originally written in 1980, but brought up to date in 2002.

Some useful websites

http://apod.nasa.gov/apod/
NASA's sponsored collection of wonderful images, one new each day.

http://background.uchicago.edu/~whu/beginners/introduction.html
Wayne Hu's webpages on the CMB.

http://firstgalaxies.ucolick.org/galaxies.html
Lick Observatory site dedicated to work on very high redshift (young) galaxies.

http://hubblesite.org/
NASA-sponsored site for all things related to the Hubble Space Telescope.

http://lambda.gsfc.nasa.gov/
NASA's archive for all things related to the CMB. This site includes the online CMBFAST computer simulation of the Universe.

http://map.gsfc.nasa.gov/
NASA's site for the Wilkinson Microwave Anisotropy Probe satellite, including a Cosmology 101 tutorial.

http://superstringtheory.com/
Quite extensive tutorial on strings, black holes, and cosmology.

http://www.astro.ucla.edu/~wright/cosmolog.htm
Ned Wright's cosmology tutorial site.

http://www.astro.virginia.edu/~dmw8f/BBA_web/index_frames.html
My own account of Big Bang acoustics.

http://www.jwst.nasa.gov/index.html
NASA's website for Hubble's successor, the James Webb Space Telescope.

http://www.particleadventure.org/
Similar to the Universe adventure is this particle adventure site, created by the Lawrence Berkeley National Laboratory.

http://www.Universeadventure.org/
Contemporary Physics Education Project has provided this interactive path through cosmology.

Notes

Notes